# Fighting on Empty

# FIGHTING ON EMPTY

———◄◆►———

*How Hitler and Hirohito*
*Lost the Economic War*

ROBIN BROMBY

Fighting on Empty: How Hitler and Hirohito Lost the Economic War

First published 2014 by Highgate Publishing
P O Box 481, Edgecliff NSW 2027, Australia

www.highgatepublishing.com.au

National Library of Australia Cataloguing-in-Publication entry:

Author:   Bromby, Robin, 1942- author.

Title:    Fighting on empty : how Hitler and Hirohito lost the economic
          war / Robin Bromby.

ISBN:     9780987403858 (paperback)
          9780987403841 (ebook : Kindle)

Notes:    Includes bibliographical references and index.

Subjects: Hitler, Adolf, 1889-1945.
          Hirohito, Emperor of Japan, 1901-1989.
          World War, 1939-1945—Economic aspects—Germany.
          World War, 1939-1945—Economic aspects—Japan.
          World War, 1939-1945—Economic aspects—Italy.
          Germany—Economic conditions—20th century.
          Japan—Economic conditions—20th century.
          Italy—Economic conditions—20th century.

Dewey Number:   940.531

Typeset in Bembo 11.75/15pt by Cannon Typesetting

# Contents

*Introduction*                                                          *vii*
*Author's Note*                                                        *xvii*

**1** Last Glimpses of Peace                                              1
**2** Japan's Wasted Years                                               14
**3** The Race for Resources                                             36
**4** The Battle for Food                                                63
**5** The Advantages of Empire                                           81
**6** The Axis loses in Latin America                                   106
**7** Italy—the weakest link                                            120
**8** The slow economic strangulation of Germany                        135
**9** The Co-Prosperity Illusion                                        168
**10** Japan Feels the Squeeze                                          187
**11** China—Japan's critical failure                                   199
**12** Losing the propaganda war, too                                   215
**13** Picking up the pieces                                            227
**14** Loose Ends                                                       250

*Postscripts*                                                           263
*Bibliography*                                                          270
*Index*                                                                 284

# Introduction

HIRAKO NAKAMOTO WAS ONE of a group of Japanese schoolgirls aged between twelve and fourteen who in 1942 were sent to work in a makeshift Hiroshima aircraft factory. After standing for hours at their machines, Nakamoto recorded that the girls on the night shift were marched into a dining hall where they had supper, which was 'a bowl of weak, hot broth, usually with one string of noodle in it and a few soybeans at the bottom', she recalled. It was cold in the factory and, at nine o'clock, the girls would each be given a small cake made out of weeds; however hungry they were, though, they could not bear the taste.

Meanwhile, across the Pacific, stores in the larger American cities in 1942 began staying open at nights due to the fact that so many women were working and were left with no time to shop; until then, late night shopping had been uncommon. Reporting on this new retail experience, the *Washington Post* described the experience of 23-year-old War Department stenographer Sarah Monica who finished work at 5.15 pm, after which she went to her favourite department store and 'after stuffing down devilled crab, coleslaw, French-fried potatoes and cherry pie, she invaded the counters laden with goods to adorn and enchant'.

Thus, on one of many levels, the war was lost and won.

So many books on the Second World War—yet so few that even acknowledge, let alone address, the economic aspects of the war. There are, as a result, great blanks so far as the general reader is concerned, and most of those blanks consist of economic issues that helped decide the war. In short, none of the three Axis powers—Germany, Italy or Japan—was prepared for total war. Japan was short of raw materials, Germany gambled unsuccessfully that it would be able to acquire key commodities by conquest, the Italians were dependent on those countries which would become their enemies (including the United States) for much of the critical commodities that kept their economy running. It was, in short, a mess. And that mess doomed the Axis as much as the superior firepower of its opponents.

Economic power in Britain and the United States (along with Russian manpower and the Soviet-built industrial base) eventually decided the Second World War. Britain certainly possessed that economic power and, once the United States was engaged, there was never really any possibility other than the Allies would win. Indeed, Germany was no match for the economic strength of the British Empire even before the American entry in late 1941; notwithstanding the popular imagery of embattled, plucky little Britain, those who governed in London had many economic tools at their command. And even the Soviet Union, once it had absorbed the shock of the German invasion, was able to bring enormous industrial capacity to bear, capacity which could not be matched on the other side of the front lines.

This whole image of 'besieged, badly prepared Britain' was effectively exploded by historian David Edgerton in his book, *Britain's War Machine*. As he pointed out, Britain was Europe's largest economy and in 1942 was the greatest producer in the world of armaments (the Americans would soon take over), and the British government was 'full of experts, of scientists and economists and businessmen'. For all the country's financial strains and stresses, and the losses incurred in the Battle of the Atlantic, during the war Britain imported unprecedented quantities of fuel oil, petrol, aviation spirit

and lubricating oils. Even as early as 1939, British factories were producing aircraft at a rate greater than Germany's plants. And Britain
had a far more extensive industrial base: statistics for 1938 showed
that only eight per cent of Britons were engaged in farming, against
thirty per cent of Germany's workforce (and even then, as will be
shown in a later chapter, the latter still could not feed the country).
Even with coal Britain was ahead: in 1939 its mines produced 236.8
million tons, the German mines 187.9 million tons.

At the end of the war Britain's Ministry of Information issued a
book, *What Britain Has Done 1939–1945*. One example might do:
as part of the Eighth Army's mission to chase the Germans out of
North Africa, the British government was able to supply this force
with 120,000 lorries. These vehicles moved—daily—2,400 tonnes of
stores. In addition, the wear and tear on this fleet required 2,000 new
tyres to be available each day. The Germans, meanwhile, depended on
horses for eighty per cent of their wartime transportation. German
planning for the invasion of Britain, Operation Sealion, included a
requirement for 11,200 horses, and for that of Russia 65,000 beasts.
Between the outbreak of war in August 1939 and the Normandy
landings in June 1944, British shipyards produced 722 major naval
ships, 1,386 Mosquito naval craft and 3,636 other vessels including
landing craft.

Then along came the Americans. In April 1938 a British delegation arrived in the United States to survey aircraft manufacturing
capacity. One place visited was the Lockheed Aircraft Corporation;
even before the British turned up at the plant's gates, the Lockheed
designers had come up with a design to produce a military version
of their Super Electra airliner. It was named the Hudson, complete
with bomb bay and three machine guns. Sir Arthur Harris, later
to head Bomber Command, recorded his astonishment that the
American designers could complete a mock-up of the aircraft in just
twenty-four hours. By June, the Royal Air Force signed a contract for
250 Hudsons for delivery by December 1939.

Between 1939 and 1944 the total volume of production in the
United States nearly doubled. This meant Americans were able to

avoid any savage reduction in the civilian standard of living while, at the same, meeting all the needs of the war effort. As the League of Nations reported in its 1945 summary of the war, 'in other words, the United States in 1944 had, in effect, two economies: one for war, and the other for the civilian community and each of them was practically equal in size to the entire pre-war economy'. While the Americans were not quite as prepared as they should have been in 1941, quite a good deal had been done: the Roosevelt administration since 1933 had spent $1.3 billion on armaments, including 5,640 new military aircraft and 1,700 anti-aircraft guns. And while there was rationing in America, a shortage of tyres and the suspension of domestic automobile production, these and other measures certainly meant inconvenience and irritation for the civilian population but came nowhere near the suffering experience by German, Italian and Japanese civilians.

This is the story of how Germany, Italy and Japan failed because, even before the fighting began, they had each sown the seeds of their eventual and inevitable economic collapse. Some of the reasons are well documented and part of general knowledge, such as the fact Germany had woefully inadequate supplies of oil. And then you discover that in 1942 German locomotive factories were still building 119 different types of steam engines, with all the lack of economies of scale and the need to make no doubt thousands of different parts to keep all those classes in operation. In 1938 Germany's railway system was making do with 4,000 fewer locomotives than it had possessed in 1929, and 80,000 fewer railway goods wagons than nine years previously (and those that remained were often in poor condition). To move all the material needed for fortification of the western frontier the railways needed to provide 6,400 rail wagons a day in the lead-up to the war; this was managed only by transferring rolling stock that was desperately needed to haul coal in the Ruhr.

But this is also a story of both astonishing lack of forward planning—in Japan's case, almost breathtaking in the extent of its neglect—and of extraordinary misallocation of resources; this included the assassination in 1936 by army officers of the country's finance minister (who had wrought spectacular economic recovery) because he planned to make cuts to Japan's military budget.

Italy's situation was similarly astonishing. Mussolini's nation depended for most of its oil supplies on the United States or parts of the Middle East (Iraq and Iran) under British influence, it mined only fifteen per cent of its coal needs, it produced barely ten million tons a year of iron ore, meaning that its steel industry was heavily reliant on imports. Italy could not even manage to be self-sufficient in food with the agricultural sector unable to feed the population and large quantities of its grains sourced through imports. As for the Germans, apart from the human horror that was the Holocaust, the diversion of resources to genocide (including disruptions to the railway systems needed for war materiel and troop movement when critical battles were being fought) is inexplicable until you comprehend the warped Nazi theories that underpinned it and demanded the murder of millions.

But there were less frightful and almost risible misallocations.

As Germany invaded Poland and triggered the Second World War, work was continuing on one of the last improvements intended by the Nazi government for the peacetime wellbeing of the German people. That September in 1939 there were proceeding, on the Baltic island of Rügen, the last stages of the building of the 'seaside resort of the 20,000', the Prora resort. Only the festival hall and the dining areas (which could accommodate the 20,000 people) needed still to be completed. Even without those final touches, this structure had achieved the distinction of being the world's longest building.

While designed originally to accommodate 20,000 German holidaymakers at a time—the vacation stays were to be ten days long—it was never completed, and never used for German workers of the Reich for whom it had been intended. Later in the war, though, the buildings were used to house people who had their homes in Hamburg destroyed by the Allied bombing.

In 1936 Nazi's labour front leader, Robert Ley, was there for the laying of the foundation stone. He then went on to head the sponsor of this monstrous complex, the KdF—the Kraft durch Freude, or

'Strength through Joy' movement. The German people were prom-
ised the complex would be the most colossal seaside resort in the
world. Centred on a square, there was to be a huge festival hall and
cafes. A road and rail bridge was built to link Rügen to the mainland.
The accommodation blocks stretched for nearly five kilometres, containing 11,000 two-bed rooms. There was a hospital and a cinema.
Hitler's plan was to use the resort to 'stiffen the nerves' of the German
people when his intended war came; the Nazis planned a number of
resorts but only Rügen's was ever built.

There was also a large pier to link the resort with another element
of the 'Strength through Joy' plan: the KdF ocean cruise liners. The
second of these, the *E.S. Robert Ley*, had been delivered from the shipyards only on 24 March 1939—this, astonishingly, while Germany's
first, and only, aircraft carrier was only partly complete in the shipyards.
The *Graf Zeppelin* was intended to carry forty-two aircraft. It had

*The KdF resort at Rügen as it appears today. The guest rooms all faced the sea with the
halls and stairwells on the landward side.* Geoffrey Walden.

been laid down in 1936 and launched in 1938, one of two aircraft carriers planned. When the war began the carrier—still incomplete, and never used in battle—was towed to Danzig (now Gdansk) and then was scuttled by the retreating German troops in 1945.

But no such delays affected the cruise liner plans. With her older sister ship, the *Wilhelm Gustloff*, the *Robert Ley* was designed to provide pleasure voyages for the German people within the KdF programme. Hitler had attended the launching in Hamburg of *Robert Ley* on 29 March 1938, *The New York Times* reporting the event and that the ship would cruise regularly to Madeira, Portugal, the Mediterranean and the Norwegian fjords. The new vessel had completed only thirteen voyages before war broke out.

While the Germans were devoting so much time and energy to holiday plans that would never be realised, the British had been concentrating on rather more essential matters, such as assessing their raw materials and foodstuffs position—or predicament, depending on which item was being considered. A report by the Royal Institute of International Affairs, published just eight days before war was declared, noted that Britain was dependent upon imports for at least two-thirds of her food (compared with less than one-fifth in Germany's case). The deficiencies were particularly striking in the case of fats, cereals and sugar. The authors had ruled out butter from Denmark and the Baltic states and the alternative, New Zealand, was 'almost too distant'. Wheat could be obtained from Canada, Argentina and the United States, and also possibly from Australia. The report assumed continuing supplies of sugar from the Caribbean, Mexico and South Africa, and fruit from the West Indies, Central America and tropical Africa. Meat was not seen as a problem with plenty available from the Americas, although Australia and New Zealand would probably be needed for egg supplies. But there was concern about tea, all growing territories being in Ceylon, India or the Far East.

The post-war literature has weaved a misleading picture of the comparative strengths of the protagonists. There has long been a perception that German strength was so much greater than Britain's when war started out in 1939; after all, the Germans had been rearming in overdrive while, for example, the Royal Air Force during the Battle of France still flew the antiquated Fairey Battle bombers while the Royal Navy was using the antiquated biplane, the Fairey Swordfish, with its maximum speed of 229 km/h (although not so antiquated that Swordfish pilots could not sink one Italian battleship and disable the *Bismarck*).

While the Germans and Japanese faced constant squeezes on armaments and materiel, Britain managed to keep its fighting forces supplied and, on top of that, to ship enormous quantities to Russia (rubber, tin, wool, lead and jute) along with handing over to the Soviets 5,031 tanks, 4,020 vehicles, 2,463 Bren gun carriers and 1,706 motorcycles. In addition, 4,106 fighter aircraft made in British plants were sent to the Soviet Union, as well as twelve minesweepers.

## II

It was estimated in 1945 by the League of Nations—it was still in existence, although a shadow of its former self—that in 1943 roughly one-third or more of total world production was for war purposes. At least one-half of the resources required for the world's war production seems to have come from the expansion of aggregate output, the rest being provided by reductions in consumption and in private capital outlay. In other words, after all the economic contractions and (at best) stagnation of the 1930s, the world economy enlarged by an extraordinary degree. Yet this astonishing process plays little more than an incidental role is most of the literature of the Second World War.

Millions upon millions died. Vast swathes of Asia and Europe were laid waste. Great battles were fought, extraordinary bravery shown by so many. That much we remember. But there is plenty about the Second World War that we have forgotten. The economic struggle has been kept largely in the background. It surfaces from time to time—rationing in Britain, starvation in occupied countries, the bombing of German industry, the relocation by the Russians of

factories in the Urals, these and many others make cameo appear-
ances throughout the body of war literature.

But this part of the story deserves better.

So (apart from passing references), no Battle of Britain here.
No Stalingrad. No Hiroshima. No Pearl Harbor. No Battle of the
Atlantic. No Iwo Jima. No convoys to Murmansk. No D-Day. No
Bomber Command. No Blitz. No Panzer divisions. No fall of Berlin.
No Churchill, Roosevelt, Hitler or Stalin. (Or, at least, not much.) In
this book, the fighting is largely in the category of 'noises off'.

Nor will you read here stirring tales of bravery in battle, no
breathtaking feats in the air, on land or sea, and no extraordinary
technological achievements such as Ultra or bouncing bombs.

If anyone has made an industry of Second World War publishing,
then it certainly is the British. Hence a great swathe of the available
literature is, understandably, Anglo-centric: the Battle of Britain, the
Battle of the Atlantic, D-Day, the blitz, children being packed off to
the countryside. But, as Max Hastings has demonstrated graphically
in his magisterial study of the war, Britain did not have such a bad
time of it if you reckon the cost in terms of lives lost; and, indeed,
he paints a rather grim picture of Britain's military management and
patchy performance of British troops. As he shows, the people of the
Soviet Union had a far worse time of it and played a much greater
price on all levels. Indeed, New Zealand forces lost a great many
more men on a per capita of population basis than did Britain's.
New Zealand, at least until the middle of 1944, maintained a higher
percentage of its manpower in the armed forces than any other of the
Allies except the Soviet Union. The extraordinary economic contribution of India and other parts of the empire have been neglected
outside academic journals.

The Americans have not been quite so preoccupied with the
Second World War as has their trans-Atlantic ally but, again understandably, their literature focuses on the United States effort. Various
other countries have their own bodies of work, but inevitably these
focus on their own parts in the war—the Australians and New
Zealanders on the desert war in North Africa in which their forces

played such a significant part, for example. Even the Italians have a substantial war literature; indeed, there's a bookshop in Bologna I came across that is largely entirely devoted to Italian-language books on the conflict and the Italian units.

Yet the neglect of certain aspects of the Second World War goes further. As one Nigerian academic, A. Olorunfemi, noted in a 1980 article in the International Journal of African Historical Studies, while the contribution made in men, money and materiel by Britain's West African colonies during the war was considerable, 'few references have been paid so far to the price paid by most of these countries for their sacrifices in aid of the mother country'. In fact, he said, there had been (by 1980) no systematic attempt to examine in depth how these contributions disrupted West African economies during the war. Little has been done since.

For so tumultuous an event, and one on which a body of literature far greater than any other subject in the history of the world has been built, so many stories of this war have been forgotten and neglected. Certainly, the plight of farmers in Nigeria, housewives in Chungking, factory workers in French India, businessmen everywhere, the people who built a war industry in India almost from scratch—these are among the many stories which have not had a torch shone on them in decades.

So much non-military history has been neglected. This is an attempt to rectify a little of that neglect.

# Author's Note

As to stylistic matters, I have generally kept with measurements of the time, particularly referring to tons rather than tonnes (although the latter will appear if the document specified 'metric tons'). Distances, though, are in kilometres: even readers in countries still using miles and feet can cope with kilometres and centimetres.

Names are rendered as they were then, particularly in regard to China. No reader will be left wondering in relation to 'Chungking' rather than the present 'Chongqing', but for lesser known places the modern spelling has been added in parentheses. I have preferred to avoid the situation found in so many modern titles where an author sticks with 'Chiang Kai-shek', for example, because to render his name in Pinyin (Jiang Jieshi) would leave readers scratching their heads, but then the authors use the modern version for lesser-known individuals, sending the reader to Google to discover the Wade-Giles version of the name by which they were known at the time. Wade-Giles it is, at least in this text; the Pinyin version of the name is added parenthetically where needed.

The country known today as Indonesia was, during its years as a Dutch colony, variously described as the Dutch East Indies, the Netherlands East Indies, or just Netherlands Indies. As the last mentioned was the name that appeared on the postage stamps, I have opted for that name.

Unless otherwise specified, all monetary sums are in U.S. dollars.

On the subject of sources, I have avoided notes within the text. When I am quoting someone else's research, I mention their name in the text and the full details are spelled out chapter-by-chapter in the Notes section at the back of the book.

Readers will also notice that the content is probably weighted toward the Asian theatre. That was deliberate because it has received far less attention than the European one.

# 1.

# Last Glimpses of Peace

ON 26 AUGUST 1939 the Sydney stock exchange rose sharply from the previous day (when shares had fallen to the year's low). Under the headline 'Vigorous Rally', the *Sydney Morning Herald's* financial editor attributed the boost that Saturday morning to Friday's rises in New York and London, which in turn appeared to be based on the fact 'that the visit by the British Ambassador [in Berlin, Sir Neville Henderson] to Herr Hitler was at the latter's request'. Berlin's stock market had closed on the Friday 'in almost boom conditions', the newspaper reported. Henderson had been called in by Hitler to discuss the Polish issue, the Germans looking for some way to achieve their invasion without provoking a war with Britain and France. The markets were grasping at straws that war might be averted.

The next week, as war seemed inevitable, Wall Street held its nerve. Investors remembered the period 1914-1917 when, before entering the Great War, United States industrial companies thrived on equipment orders from the belligerents. Those companies went down in history as the 'war bride' stocks. On 4 September 1939, the day after Britain and France had declared war, investors in the United States kept moving into the market, expecting a repeat of war order bounties. In the first eight days of the month, the Dow

Jones Industrial Average rose by fifteen per cent; on one day in late August, the index rose 7.3 per cent. It would not last, of course; the peak of 155 points reached in mid-September would not be seen again until 1945, and the 1939-1942 bear market was soon to begin. On 1 September 1939 the Federal Reserve Bank reported the arrival in the United States of $30.47 million worth of foreign gold, $24.79 million in New York from Britain and the balance arriving in San Francisco from Japan.

Meanwhile, John H. White, in a cable from Buenos Aires to *The New York Times*, reported that South Americans were seeing the looming war as a boost for their own prosperity. Argentine business-men welcomed war, he reported. 'They make no attempt to hide their eagerness'. Argentina and Uruguay were expecting to reap the rewards of shipping food to the warring nations, while Bolivia eagerly expected to move unsold stocks of tin and copper; the other items that were exciting South Americans were their exports of manganese (Brazil), copper (Peru and Bolivia), petroleum (Peru and Venezuela), nitrates (Chile) and minerals generally (Colombia).

If you had not been following the news, so much seemed so normal as August 1939 came and went, and then September arrived.

On 2 September 1939—the day before the war in Europe began—*The Times* newspaper was still listing the frequencies and programmes being transmitted from German radio stations at Cologne (including its 10.15 pm news in English), Königsberg (now Kaliningrad in Russia), Munich and Vienna, along with broadcasts from the Italian stations; that night, Milan was broadcasting the opera *Turandot*. Then, in Monday's edition, the powerful 150,000 watt long-wave Deutschlandsender station in Berlin was listed in *The Times* as having dance music at 8.15 pm followed by a program 'Youth on the Frontier'. Military band music was featuring on the Cologne station, while Rome No. 1 had news in English at 7.18 pm. On the very pow-erful 200,000 watt Vienna station, Austrian folk music was followed at 9.00 pm by a Mozart serenade concert from the Salzburg Festival.

Between 23 August and 2 September—the last day being on very eve of the Second World War—at the great Olympia exhibition centre in London there was being held the annual Radiolympia, showing off the latest wireless sets, loudspeakers, aerials and anti-interference devices and the infant medium of television. One telecast from Olympia featured the debut singing and dancing performance of an eleven-year-old up-and-comer, Bruce Forsyth. ('Brucie', as his affectionately known and now a television legend in Britain and Sir Bruce, was still performing in 2014 as this was written.)

With the Second World War just a few weeks away, life was going on much as usual for many as August went by. In Bayreuth there was the Richard Wagner festival, although the 1939 performance was the last Hitler would attend; the Nazis had banned the performance of *Parsifal*, presumably because of its underlying themes of pacifism (although there is still disagreement about the motivation behind the ban). In mid-August the management of the Salzburg Festival announced the event would close on 31 August instead of the originally planned 8 September. The reason given at the time was that the Vienna Philharmonic Orchestra was required to perform at the Nuremberg Party Convention to be held in the first week of September (which had already been cancelled in a secret order from Berlin). Hitler had been in the Bayreuth audience in early August for two Mozart operas, *Don Giovanni* on 8 August and *The Abduction from the Seraglio* on 14 August. The government had ordered the festival organisers to remove from the entrance foyers frescoes by Anton Faistauer—an Austrian pioneer of modern painting—and to take instructions on all design matters from Benno von Arent, a Nazi Party member.

In Dublin on the first weekend of September the Zoological Society announced it had just taken delivery of two leopard cubs and one Himalayan bear cub presented by Lieutenant Fallon of Punjab, India; a giant squirrel had been delivered from a Mr Phillips in Ceylon while other contributors to the zoo had given hawks, fifty-six frogs, three eels and four rock fish. In Australia, British scientists were unsuccessful with attempts to use germs to kill rabbits. More than 2,000 rabbits were fenced inside a ninety acre area and some deliberately infected; however they failed to pass the infection on to the others (a failure that would be rectified in 1950 with the introduction of the myxoma virus, the world's first biological control of animal pests). Doctors in Philadelphia were experimenting with placing women cancer patients, after they had been given sleeping sedatives, into refrigerated containers in the hope that would halt the growth of their cancer cells.

Frank Waitt, advisory engineer to the Sydney Water Board, was about to leave for the United States to study water supply systems ahead of the board's proceeding with a new scheme, the Warragamba, to meet Sydney's needs for years to come (as it did; the Warragamba Dam remains the city's chief water source). On the day before war began, in New Zealand the Fijian rugby union team had defeated Auckland by seventeen points to eleven. The *Auckland Star* reported that the Auckland back line players (the men whose job it was to make the running plays) were thwarted by the solid tackling of the Fijians and 'were nonplussed by the solidness of the native back-line, which stood close up and raced up swiftly to check attacks'.

The last peacetime edition of the British popular photojournalism magazine *Picture Post* was not totally preoccupied with British matters (although the first feature was a long photo essay on football). One of its leading features was a report on discontent under the Raj. Bombay was no longer worth living in, according to European expatriates there. The reason? The headline told it all: 'Not a Drink in Bombay'. The magazine was reporting the introduction of prohibition, forbidding most the then million and a half people of the Indian city from imbibing. It was not so bad as the grumbling expatriates made it out to be: if you were a European, actually, you could still get a drink, although not in such lavish quantities than those to which many had been accustomed. Seven units were allowed per month, one unit being equivalent to a bottle of whisky, brandy or gin, or three bottles of wine, sherry or vermouth, or nine bottles of beer or stout. The upshot was that new taxes had to be levied on property owners to replace the million pounds in excise duty foregone. The blame for the dreaded prohibition was all laid at the feet of Mahatma Gandhi, who had campaigned for this for twenty years. In addition to the existing 4,500 men of the Bombay police force, another 1,000 had been recruited to enforce the new liquor law. As an alternative to liquor shops, applications were being received for facilities to open milk bars.

One part of the British Empire still not at war for the first week of what would become the Second World War was Canada. Prime Minister Mackenzie King said he would not commit the country to war without the authority of parliament. This reluctance to follow the prime ministers of Australia and New Zealand, who had declared war on their own authority immediately after Britain's declaration, may have been influenced by the fact that Canada had the previous year been riven by disagreement over whether to get involved in what was obviously shaping up to be another European war. King himself had pacifist leanings and, in the process of accepting the need to support Britain, knew he had to move public opinion to a similar conversion. On 10 September the parliament sanctioned the move with dissent only from James Woodsworth, a pacifist ex-Methodist minister who had formed the Commonwealth Co-operative Federation (although all the other CCF parliamentarians voted for war), and a few French-Canadian members of King's Liberal Party who remained isolationist.

South of the forty-ninth parallel, the Americans remained insouciant regarding the need for industrial mobilisation, even though the first report preparing for this had been issued in 1922 (under the Harding administration, the planners in Washington back then anxious not to get caught unprepared as they had been in the First World War). Unfortunately, the military authorities adapted the attitude that it was inconceivable that they should mobilise industry until war was actually declared. Four agencies to organise for such an eventuality had been set up in the inter-war years, but each had been a failure. Symptomatic of this state of mind was the minimal effort put into the 1939 Industrial Mobilisation Plan: it was all of eighteen pages. The 1936 plan had stretched to seventy-five pages but its failing was that, while the question of military supplies was addressed, the needs of the civilian population in wartime were not. Even after the defeat of Poland no one in Washington bothered to start planning for industrial mobilisation. It took the defeat of Belgium, the Netherlands and France in 1940 to concentrate minds in Washington.

In Zurich in this month of 1939 there would open the Swiss National Festival (although that was overshadowed by the mobilising of 430,000 Swiss troops on 1 September). In Scotland, the Highland Games were due to be held. There was the Shakespearean festival at Stratford-on-Avon. Budapest had on August 20 celebrated St Stephen's Day (to honour the king born in 1038 who would go on to create the Hungarian nation), while on 16 August there was the famous Palio de Siena horse race in that Italian town. If religious festivals were a tourist's desire, then in Breton they were about to hold the 'pardons', ceremonies dating from medieval times in which communities would celebrate a local saint complete with masses, torchlight processions and capped off by singing and dancing fuelled by large quantities of wine.

In South America, Peru was well advanced with organising its first national census since 1876. (It finally was set down for 9 June 1940. It counted the population at 7,023,111.)

Countries that would soon be swamped by Axis forces were in August still issuing new stamps reflecting their national pride. Belgium put out a set of five on the occasion of the International Railway Congress meeting in Belgium (which included delegations from China and New Zealand); Bulgarians could buy a new issue in five denominations commemorating a gymnastic congress while Hungary issued stamps from two forint to 20 forint face values for the girl scout congress held in Godollo national park. That week also, the late Marshall Pilsudski (Poland's chief of state after it gained independence in 1918) featured on two new Polish stamps along with marching columns of troops; the third stamp in the series bore the image of Marshal Edward Smigly-Rydz, head of the country's armed forces and who would on 1 September face the impossible task of resisting the German invasion.

On the very day that Neville Chamberlain broadcast to the British people with the news that war had been declared on Germany, the *Los Angeles Times* told its readers of the Japanese goodwill aviators due to land at Burbank airfield at 10.00 am that morning, after which the flyers would proceed by motorcade to the 'Japanese section of the city'. Prominent members of the local Japanese community were to welcome the flyers at a banquet in the evening. Not everyone was so welcoming: the United Korean Society, the China Aid Council and the American League for Peace and Democracy had prepared banners protesting at the destruction by Japanese troops of almost half of China's cultural institutions, hundreds of hospitals and missions and orphanages. Other banners reminded the local crowds of the 75,000 bombs dropped on Chinese cities, the 36,000 dead civilians and the destruction of 115,000 homes.

## II

Britain's Colonial Office in London had to keep operating even with the nation making preparations for war in Europe; as it would turn out, the resources of the empire were critical over the next six years. In mid-August 1939, Major William Scupham was appointed administrative secretary of the Tanganyika territory; David Willoughby Saunders-Jones was being told he was off to Nyasaland (now Malawi) as administrative officer, his previous colonial experience having been six years in the Indian Army followed by seventeen years in Zanzibar; William Marchant was appointed resident commissioner of the British Solomon Islands Protectorate (and would later oversee the evacuation of Europeans ahead of the invading Japanese and then help set up the coast-watcher system that serve Allied navies so well. In 1943, he would move to Kenya as chief native commissioner.)

There were plenty of more minor appointments. Mr T.A Strong was to be the new conservator of forests for Ceylon, Dr H.M. Nevin was going to Penang as pathologist while V.H. Brooke, formerly a sergeant in the Palestine Police Force, had been promoted to assistant superintendent of police in Trinidad.

Governors were required to issue many ordinances to do with the war—updating official secrets provisions, providing for stockpiling of foodstuffs, tightening immigration controls into various colonies, increasing censorship (especially of telegraphic traffic) and setting up military units (including the Seychelles Defence Force and the Royal West Africa Frontier Force in The Gambia). Ceylon also introduced rules for blackouts if the need arose.

### III

American shipping lines decided in August 1939 to boost their numbers on the trans-Atlantic run by announcing reduced fares in tourist and third classes, with prices being slashed by twenty per cent. It was confidently predicted that many people would take advantage of this to travel to Europe by the end of the year and into 1940. Even seven months after the European phase of the war was under way, Americans still believed life would go as normal. Douglas Malcolm of American Express in April 1940 predicted that year to see a record amount of travel. He singled out national parks in the United States and Canada along with mountain, beach and ranch vacations; the $250 million that formerly had been spent on travel 'overseas' (by which he presumably meant travel to Europe) would see added sales for merchants and hotel keepers 'from the Yukon to the Amazon, from Panama to Punta Arenas,' he told the *Chicago Tribune*. The General Federation of Women's Club was at the same time organising a good neighbour tour of South America lasting fifty-two days, travelling by the liner *Santa Clara* from New York to Santiago, Chile, then by train to Buenos Aires where they would join the vessel *Uruguay* for the cruise homebound. Alaska's tourism industry was expecting good summer business because holidaymakers could travel in neutral waters by ships flying the stars and stripes.

Meanwhile, the Japanese shipping lines were putting the best possible face on an increasingly tense relationship across the Pacific. Back in July 1939, the United States government had abrogated the 1911 trade treaty with Japan in protest at the latter's actions in China and destruction of American property there. This allowed Washington to impose embargoes, although by early 1940 these had not actually been brought into effect. So the Japanese shipping lines servicing American ports were carrying on with business as usual. In fact, officials of the NYK (or Japan Mail) Line were reported as planning a 'tremendous' expansion programme according to *The New York Times*. Three of its new passenger vessels were due to begin sailings to California in March; these ships had been designed to run the route to Europe through the Suez Canal but the outbreak of war there put paid to those plans. The first sailing was to be on the 16,500 tonne *Nitta Maru*, leaving Yokohama on 28 March for San Francisco with space for 127 first class passengers, along with eighty-eight in second and seventy in third class. (The *Nitta Maru* was later converted to an escort carrier and then sunk in 1943 by an American submarine.)

Ten days before war was declared, the *Los Angeles Times* reported that 'marine men' looked upon as a good omen for peace the fact that not one German liner had interrupted its normal sailing schedule. Contrast that, the paper said, with the Munich crisis of the previous year when ships flying the swastika were ordered to turn for home. While, at the end of August 1939, the *Europa* was in mid-Atlantic on its way to Germany and the *Hamburg* was one day out of New York also headed for home, four vessels (*Bremen*, *Hansa*, *New York* and *St Louis*) were heading for New York, the *Deutschland* was berthed at Southampton and the *Columbus* was cruising in the West Indies.

Much comfort was also being taken from the fact that Italian lines, on the Monday following the French and British declarations of war against Germany, had announced plans for new trans-Atlantic passenger sailings. But within days, the Italian shipping companies had hiked their fares by up to sixty-six per cent as insurance premiums soared. Not only that, but there was an expectation that, while the ships would be full travelling to New York, there would be few passengers wanting to travel to Europe now that war had begun. Within a month, the Italian line announced it was withdrawing the *Rex* and the *Conte di Savoia* due to the 'ruinous cost of operation'. The *Conte di Savoia*, along with two other liners, *Exochordia* and *Noordam*, made their final New York departure for Italian ports in late November. (Aboard the Conte di Savoia was a Major Matahiko Yosita of the Japanese army who told reporters he travelling to Rome for 'sightseeing'.)

By 1940, of course, relaxing crossings by trans-Atlantic liners were but a memory and, while the United States was not yet at war, many Americans were beginning to realise the potential inevitability of growing involvement. This could be seen at that year's New York's World's Fair. While the fair's first season in 1939 was dominated by what no doubt the average American would have called 'new-fangled' technology, by its second season in 1940 the fair had been rededicated to peace and freedom. As historian Marco Duranti wrote, it was a switch from utopia to nostalgia. In essence, it was no longer possible in 1940 to pursue the escapism of the previous 'world of tomorrow' theme: a public that had been traumatised by the First World War and then the Great Depression was very quickly jerked back to reality by the time Germany was marching across Europe. *The New Yorker* considered in 1940 that the fair should change its title from 'The World of Tomorrow' to 'The World of Yesterday', so apparent was what it saw as the forced gaiety of the crowds attending in the fair's second year. By June 1940 changes at the fair made the war inescapable for all those visiting it: there were the activities of the Polish-American community, the pavilions representing France and Belgium, the British pavilion showing newsreels of latest developments, plus the exhibitions staged by other belligerents (Australia, New Zealand, Canada, Denmark, Norway, Czechoslovakia, Luxembourg, and Finland). Meanwhile, Italy spent five million dollars on its exhibit at the fair, described as one of the gaudiest pavilions (it included a sixty-one metre tower supporting a statue of the goddess Roma). Japan's pavilion, in the shape of a giant Shinto shrine, included a Liberty Bell replica with a diamond and pearl frame. Germany was not represented.

And, within a month of the 1940 season beginning at the New York fair, and while it was still in full swing, the Italian government announced that the 1942 World's Fair, which was to have been staged in Rome in 1942, was postponed indefinitely. As Camille M.

Cianfarra filed from Rome to *The New York Times*, it was a sure sign
that Italy was planning to enter the European war. 'Only very important reasons could have prompted Premier Mussolini to postpone
an event that was to glorify before the world twenty years of fascist
achievement,' she wrote.

Germany's free hand in eastern Europe as a result of its newly-
signed non-aggression pact with the Union of Soviet Socialist
Republics had shocked the leaders in Tokyo. They feared that Stalin,
no longer having to reinforce in the west against the Nazis, would
move his armies to intervene in Japanese-held areas of China. August
1939 ended with reports that new Japanese troops were pouring
into the puppet state of Manchukuo to reinforce the 500,000-strong
Kwantung Army and protect the 1,600 km-long border with the
Soviet Union. The non-aggression pact inflamed Japanese opinion
and Associated Press reported that Japanese gendarmes stationed
in the occupied city Tientsin (now Tianjin) had begun slapping
German residents with the result that many of them decided to
leave. AP noted that swastika emblems flown on cars owned by
Germans in Shanghai and Peking had also disappeared. Meanwhile,
Japanese authorities in Tientsin ordered there be no further anti-
British demonstrations (both the British and French concessions
having been under siege by Japanese forces). Meanwhile, plans were
being made to withdraw the Seaforth Highlanders from Shanghai
and the cruiser *Birmingham* from Tsingtao (now Qingdao) to join
the war against Germany.

In San Jose, the German consulate was advising all Germans
living in Costa Rica to voice their support for the Nazi cause and to
be ready to serve the Reich.

## IV

Four months into war, the British motor industry was still doing
quite nicely. The Society of Motor Manufacturers announced in
mid-January 1940 that factories had exported three hundred vehicles
on each working day of November 1939, which was up twenty-six

per cent on the number for November 1938. And, during the period
September to November 1939, records had been set for exports to
Australia, India (including Burma), Portugal and Uruguay. Other
substantial markets were Argentina, Ceylon, West Indies, Thailand,
Eire, Malaya, Australia, Canada and South Africa. However, there
had been serious reductions in exports to Scandinavia and to New
Zealand, although the latter country had banned imports of motor
vehicles from all countries other than the United Kingdom.

For a while as the war entered its early months, things so far
as the British were concerned seemed (almost) normal. While
Germany was increasingly cut off from much of the world, Britain
for all its troubles could still interact with many of those countries
neither allied with nor occupied by the Nazis or Italians. Almost a
year after the European war began, the aviation editor of the *Christian
Science Monitor* was reporting it was still possible to send passengers,
letters and packages by air 'over most of the earth's surface'. One
of the most important exchange points for mails between neutrals
and belligerents, and between Europe and North America, was
Lisbon. Pan American Airways had three services a week to the
Portuguese capital, the only place open to United States aircraft due
to regulations prohibiting the airline's entry to the combat zones in
Europe. Letters and parcels from North America were separated in
Lisbon, those for Britain being put aboard twice-weekly flights to
London; Iberia aircraft linked Lisbon with Madrid and Barcelona;
Ala Littoria, the Italian airline, flew thrice weekly to and from Rome,
which linked there with Lufthansa flights, allowing mail from North
America finally to reach German recipients. For those awaiting letters
and parcels in North Africa, Aero Portuguesa operated from Lisbon
to Tangier.

In Asia, British commercial flights reached Bangkok, Rangoon,
Karachi, Hanoi and Hong Kong; southwards, Singapore, Batavia
(Jakarta) and Sydney were linked by air. At Hong Kong, passengers
and mail could continue across the Pacific on Pan American aero-
planes that travelled via Manila, Wake, Guam and Midway Island to
San Francisco and Honolulu. Eastwards, China National Airways

took off from the British colony for Kweilin (now Guilin) and Chungking, thence to domestic flights and a link also to Hanoi.

In November 1941 the United States post office announced an airmail service between the U.S. and Africa, the letters travelling across the Atlantic from Miami to Leopoldville, the capital of the Belgian Congo (with stops along the way in Puerto Rico, Trinidad, Brazil, Gambia and Nigeria before reaching the Belgian territory). In South America, Pan American provided the air links along the west coast to Colombia and Chile.

European airline services were among the first commercial casualties of the war. Several routes had connected Europe to South America with scheduled flights by government-owned airlines. Air France, Lufthansa and the Italian LATI (Linee Aeree Transcontinentali Italiane) had operated regular services to Brazil. Lufthansa services ceased immediately upon the outbreak of war; Air France continued to operate to Brazil until Paris fell. LATI, as will be related in Chapter Six, would face its own day of reckoning.

Of course, as the war progressed communications faced severe disruptions. *The Economist* reported in mid-1941 that an airmail letter from London to Buenos Aires usually went first to Lisbon, thence in the Pan American clipper to New York, and then along the South American coast by connecting Pan Am services. But by this time such transit times stretched out from one week to three; worse, sometimes surface mail was reaching London before air mail posted on the same day in the Argentine capital.

But, by this time, most people around the world had more to worry about than the speed of postal services.

# 2.

# Japan's Wasted Years

TEN MONTHS BEFORE IT was to bomb Pearl Harbor, Japan's economic planning was still in disarray, astonishing considering how difficult the struggle would be against the financial and resources power of the United States. Shipping was just one of the problems. The Japanese military required so many vessels to maintain supplies for its army in China—in the first half of 1939 some 1.6 million gross tons, about thirty per cent of the Japanese merchant fleet, was monopolised by the military to ship troops and equipment to the Chinese theatre and also supplying Japanese forces up the Yangtze River —that it left Tokyo with insufficient numbers of merchantmen to maintain its foreign trade and, consequently, maintain its earnings of foreign exchange. But this foreign exchange imperative, and the need to conserve spending as the scope of the war effort widened, also worked against chartering of foreign vessels to fill its trade gap.

Japan as an island nation faced a similar marine logistics challenge to that of Britain, yet its merchant marine was less than one-third the size of Britain's. Nor could Japan turn to foreign-flagged vessels to help: the numbers of those plying routes across the Pacific was diminishing due to the war in Europe. The British had declared

Danish vessels to be belligerents because of Copenhagen's decision not to resist the Wehrmacht, and the U.S. took more than thirty of those ships into protective custody. Italian and German cargo ships also were withdrawn or seized. The Royal Navy's blockade of German-occupied France also meant French vessels were bottled up in their home ports, or unable to reach them. In February 1941 it was reported by the *Christian Science Monitor* that a shortage of shipping was impeding timber exports from the Canadian province of British Columbia. Japanese companies still owned some 500 million feet of standing timber at the northern end of Vancouver Island but their shortages of money and ships was expected to lead them to sell off the timber into the domestic market.

How do you contemplate becoming part of a world war when you do not even have enough electric power to support your domestic manufacturing industry? Well, apparently, you could if you were running Japan in 1939. Power rationing had to be introduced in 1939 when chemical and metallurgical plants were not able to get enough electricity. The shortages were caused by insufficient coal supplies. In late January 1940, Associated Press reported to the world that industrial centres in fourteen Japanese prefectures were at a standstill apart from those plants producing urgently needed supplies for the front in China. Coal shortages had meant power plants in Osaka, Kobe, Kyoto and elsewhere had shut down, leaving three million factory workers idle. Even when mines could produce the coal, they often faced delays moving it to customers due to the fact that Japanese railways and merchant shipping was already congested shifting needed supplies for the war in China. Even electric clocks in railway stations were switched off. (It did not help, either, that Japan's hydro-electric stations were operating far below capacity due to drought.)

At the time, American commentators were watching, astounded, at Japan's coal problems, with production in 1939 seen by them as having fallen about fifteen per cent on the 1938 output figure. That year the newspaper *Japan Advertiser* told its readers that one coal mine had been closed because its machinery had been idled due to a broken bearing, and the company was still waiting to get a replacement part. There was also intense competition for what mining machinery was being made as Japanese companies producing coal in Manchukuo and North China were also demanding equipment. The coal supply situation just before Japan went to war with the United States was further exacerbated by the fact that the mines in Manchukuo were not producing sufficient quantities to meet industrial needs in areas of China controlled by Japan, let alone to ship much to Japan.

Then there was the critical matter of oil supply.

Japan had long depended for its much of its oil on the very country—the United States—that it was soon to go to war with. But, while Washington waited until well into 1941 to place an embargo on Japanese purchases of its oil products, the tightening had begun a year earlier. In mid-1940 the Roosevelt administration added petroleum to the list of items that required an export licence. Japan was so vulnerable: its refineries depended on American crude being available, some four million tons (31.2 million barrels) a year coming from California alone. But its import dependence also covered kerosene, lubricating oils, asphalt, petroleum jelly and paraffin wax. In 1938, Japan's own oil fields yielded 350,000 tons of crude, equivalent to less than a tenth of its imports. But there was also crude available from Japanese-owned fields on Sakhalin Island, the southern half of which was under Tokyo's control (having been won in the Russo-Japanese war of 1905).

Unlike Germany, Japan failed in its attempts to produce oil from coal; while the Germans had their own coal fields, Japanese had to import most of their coal, the fundamental flaw in any synthetic petrol plan. Japan's continuing attacks in China and moves in Indochina were the prime cause of the United States cutting off all oil shipments to Japan in 1941. This in turn made it imperative for Tokyo to seize the Netherlands Indies, the greatest source then of oil in East Asia.

By 1938 another problem had emerged: Japan's foreign exchange shortages were hampering its ability to pay for oil it needed to buy from California and Mexico. That year the government in Tokyo decreed that about a quarter of petrol had to have five per cent alcohol content, with Korea and Manchukuo expected to adopt the same requirement by 1939. There was a target set of eventually having the alcohol content at twenty per cent but, even at the time, it was conceded it would take five years to achieve that. The alcohol was to come from sugar cane produced in Japan and from cane and sweet potato in Formosa (now Taiwan). In 1937 the Seven Year Plan had called for Japan to be self-sufficient for two-thirds of its motor fuel needs and forty-five per cent of its heavy oil requirements. That year Japan, by chemical processes, was able to produce just 2,000 tons of petrol.

Not that oil planning was totally remiss. In December 1941, Japan had twenty-one oil refineries, and a reserve of forty-three million barrels had been built up over the previous ten years. But, even before Pearl Harbor, Japanese military planners realised their forces in China were consuming far more oil than expected. The embargo which would be placed by President Franklin D. Roosevelt on oil exports to Japan would wipe out eighty per cent of Japan's oil imports. Yet, as Stephen Wolborsky enumerated in a paper for the Air University in Alabama, Japanese planners had to provide fuel to 7,500 land-based aircraft, 1,180 tanks, 81,000 other military vehicles, ten aircraft carriers, ten battleships, thirty-three cruisers, 111 destroyers, sixty-four submarines, and 500 carrier-based aircraft. The secret, therefore, had to be to use this hardware for a quick and decisive victory against the United States; any war of attrition would be lost. Hence Pearl Harbor.

A week after the war in the Pacific began, *Time* on 15 December 1941 listed Japan's resources as it understood them. The new enemy had enough oil for fifteen months of war. The United States had supplied 8.2 million tons of scrap iron between 1936 and 1940 but none in 1941. Japan was heavily reliant for aluminium on supplies from Canada, on copper for imports from the United States and Chile; and most of the machine tools Japan needed for her war industry had been coming from Britain, the United States and Germany.

As Barton Biggs details in his 2010 study of war psychology, in 1941 a Colonel Iwakura, a Japanese logistics expert, briefed senior military figures in a style described as cautionary. He contrasted Japanese motor car output (48,000 in 1940) with that of the United States that year (4.5 million); moreover Japanese munitions output was one-tenth that of America's, its coal production one-thirteenth.

And most of Japan's scientific information came from the United States and Britain, its research institutes relying on co-operation with Western scientists. Even as late as 1940 a Japanese physicist was being give a guide tour of American nuclear laboratories. Ben-Ami Shillony, in the mid-1980s professor of Japanese history at the Hebrew University of Jerusalem, noted that from 1941 Japan looked to Germany to replace its contacts with the Anglo-Saxon nations. There were three problems with that: Germany was hardly in a position to forego equipment it needed desperately for itself; there were the logistical problems of shipping anything to Japan after the invasion of the Soviet Union and the loss of the use of the Trans-Siberian Railway; and, as Shillony notes, the Germans had a basic distrust of the Japanese. So when they did manage to ship radar, ultrasonics, and infra-red apparatus, the equipment was usually of older models. Suddenly Japan had to gear-up its own scientific facilities; between 1939 and 1945 it established twenty-eight new research institutes. Moreover, Japanese university students were in the majority taking courses in law, economics or the liberal arts. With the war on, there were not enough doctors or engineers. It was not until 1942 that a science bureau was set up. It was only in 1939 that a new university was opened (in Nagoya) with focus on teaching of medicine, science and engineering.

In February 1941 the journal *Oriental Economist*, then published in Tokyo and battling to maintain its reputation for independence (these days it is published out of New York), was confident that the Japanese economy, while severely strained, would be able to surmount that strain. It said that 'difficulties will increase but, provided the government can avoid faulty measures, Japan can continue to carry on hostilities on the present scale for several years more'. The key words there are 'present scale', which essentially meant the war in China, and which term presumably precluded taking on the United States and the British Empire.

But, as a correspondent in Tokyo reported to *The Times*, the government's budget had increased five-fold over the previous five years. The 1935-36 budget had been balanced; the 1941–42 budget would be laden with debt of twenty-five billion yen, compared with tax revenue in the previous year of 3.27 billion yen. 'How long Japan can continue to absorb these enormous loan issues is a speculative question', wrote the correspondent in February 1941. As it would turn out, absorbing that debt would become an even greater problem. The Bank of Japan was keeping the money printing machines busy to maintain liquidity in the economy.

But, as the Americans say, here's the thing: Japan's economy had been weakened by almost a decade of economic and trade problems. Not only did Japan suffer from the same economic contractions that afflicted the globe during the Great Depression, but its trade had been badly damaged as protectionist measures were implemented by many of its trading partners. More than forty countries raised tariffs on Japanese goods. As the 1930s began, the Japanese pinned great hopes on textile exports to earn much needed dollars. In the late 1920s, the cotton piece goods sector was a vital part of Japanese manufacturing, employing around three million people. By the 1930s it was the world's second largest producer of these goods (having overtaken Britain), with half the output being sent to export markets.

But those export markets were turning on Japan. In 1932 India imposed a fifty per cent import tax on Japanese cotton products; then in 1932 Jamaica imposed a tariff, mainly to protect its market for British manufacturing. Peru and Australia were among those countries alarmed by the quantities of cheap Japanese products entering their countries, the latter citing, among other items, Japanese rubber-soled bath slippers, men's braces and cotton towelling. Britain imposed quotas in 1934, then a year later Cuba slapped Japanese cotton goods with a twenty-five per cent tariff and silk and rayon with a 100 per cent impost. In 1936, President Roosevelt moved to

protect American manufacturers from cheap Japanese goods. The
one trend that did soften the blow to Japanese manufacturing was
that, from about 1934 with global recovery, the overall demand for
textiles was rising. The Japanese raised their plight at the 1937 World
Textile Conference but the situation was, in Tokyo, another argument
to justify aggression to secure monopolies in Asia.

## II

But if was unequal to the United States in oil, then Japan also faced
an inadequate merchant marine. Even before December 1941, the
growing shipping demand between Japan and China had placed a
strain on the nation's cargo shipping fleet. In 1939, even after freeing up about 500,000 gross tons of shipping from working for the
army and navy, Tokyo's merchant shipping fleet could not move from
China all the iron ore, coal and lumber that was needed by Japan; the
private shipping companies preferred to grab cargoes that yielded
higher freight rates.

Shortly after becoming involved in fighting in China, Japan first
tried regulating freight rates to strike a balance between the needs of
shipping to and from China and the country's wider trading business.
In 1940, of the tonnage entering and clearing Chinese ports, 53.5 per
cent of that was Japanese-flagged (followed by British bottoms with
17.3 per cent, American ships 5.7 per cent and Norwegian five per
cent—with 'Chinese junks' coming in fifth with 4.5 per cent of the
tonnage loaded and unloaded).

In 1940 Japanese cargo shipping lines abandoned services to
Europe while routes to Africa and India saw much lower density
of Japanese shipping space. The merchant fleet was to concentrate
its efforts within what Tokyo hoped would become the full extent
of its empire in Asia and the South Seas. The maritime authorities
authorised, effectively, the overloading of cargo ships and postponed
any scheduled docking for repairs in an effort to keep the available
fleet fully utilised. The issue was complicated by the decline from
1939 in the number of new ships built; Japanese provisioning for

war of raw materials was woeful and plagued the war effort from the moment the Americans were engaged. Moreover, as we shall see, seizing the valuable colonies of the Dutch, British and French was one thing—the inability to actually shift the raw materials back to the home islands was evident from the start of the war and just grew worse as the Japanese merchant fleet was progressively sunk by the Allies through to 1945. By early 1941, when Daniel Marx Jr was recording the Japanese situation for *Far Eastern Survey*, shipyards were getting only half they steel requested and shipbuilding times had almost doubled.

The American shipowners faced a different situation. Forced to withdraw tonnage from the European war zone and occupied countries, their vessels found plenty of work on the South American and trans-Pacific routes. With the disappearance of the European lines from the sea lanes of the Pacific, there was more than enough work. However, Washington had sold all of what were previously considered redundant vessels to the British, so the Americans, too, suddenly found they did not have enough cargo capacity.

The U.S. fleet had been dwindling since the build-up for the First World War. In 1922 there were 3,779 American-flagged vessels. By 1939 there were 2,345 ships.

To make up for the shortage, and to do so quickly, required the American shipyards to concentrate on types of vessels that lent themselves to mass production and quick assembly of those mass-produced parts. Enter the Liberty ships. Often called 'the ugly ducklings' of maritime work, the Liberty ship was based on the British tramp cargo ship of the 1920s and 1930s. ('Tramp', incidentally, did not necessarily mean run-down; the term applied to distinguish these vessels from the ocean-going liners that operated to tight schedules; tramp steamers had no schedule: their owners sent them where and when cargoes were on offer, and they became the workhorses of the ocean trade in the first half of the twentieth century.)

The Liberty ships, however, were all-welded rather than all-riveted and the first, the *SS Patrick Henry*, was launched in September 1940. In all, 2,700 such vessels would be built and each cost about

$2 million; they had around 250,000 parts and were welded into 250 ton sections for final assembly in shipyards. Their official designation was EC2-S-C1, the 'EC' standing for 'emergency cargo', the 'S' for 'steam', and C1 being the designation for that type of emergency vessel. They were armed with stern, bow and machine guns. In addition to the civilian crew employed by the private shipping companies operating these vessels, there were also navy personnel to man the guns. The holds of Liberty ships could accommodate 9,000 tons of cargo and tanks while aircraft and railway locomotives could be lashed on deck. One ship could, say, carry 2,840 jeeps.

The biggest shortcoming was the maximum speed, just eleven knots, which made the Liberty ships easy prey for U-boats (altogether around 200 would be lost during the war). Their great advantage, though, was the speed at which they could be built. It had been expected they would take 105 days from laying of the keel.

In 1944, just months away from being selected as Roosevelt's running mate in the November election, Harry S. Truman was chairing the senate's Special Committee to Investigate the National Defence Program. The committee held hearings into the performance of the Liberty ships. The senators heard that, in the rush to produce enough ships, shortcuts were often taken and shipyards were forced to employ inexperienced workers due to labour shortages, the mistakes made sometimes causing ships to fail or to crack at sea. Stresses to the hull and temperature changes of water had been problems. But improvements in welding techniques had meant the later ships were less likely to crack.

The committee voiced concern about the slow speed of the vessels. By this time, shipyards on the Pacific seaboard had stopped building the Liberty ships and turned to what were known as Victory ships or VC2s as they were known officially. They carried sixty-two merchant marine sailors and twenty-eight navy men. These vessels were much faster than Liberty ships because the steam engine had been replaced with the more modern steam turbine propulsion and the Victory ships could reach 17.5 knots. The first off the slipway was SS *United Victory*, launched in Portland in January 1944.

Altogether, 414 Victory ships (along with 117 Victory attack transports) were built.

They had their failings. But they were a lot more than the
Japanese had.

### III

Japan had a facility for shooting itself in the economic foot. In 1923,
the country suffered one of its greatest setbacks in the form of the
Great Kanto Earthquake which killed 140,000 people, flattened
700,000 homes and devastated the economies of Tokyo and
Yokohama. Then came the Showa financial crisis.

In 1927, with the Japanese Diet debating a bonds issue to clear
up the last of the bad debts generated by the earthquake, Finance
Minister Naoharu Kataoka told the parliamentarians—erroneously,
as it turned out—that the Tokyo Watanabe Bank had failed. Financial
panic ensued, but the government managed to calm the public's fears.
However, the flames of panic were re-ignited when it was disclosed
there were financial difficulties between the Bank of Taiwan (which
issued all the bank notes on that island) and the trading house Suzuki
& Co. A rescue effort by the government was torpedoed for political
reasons by the Privy Council, the cabinet resigned and the panic
spread across the entirety of Japan. It was only the injection of public
funds and rapid introduction of banking reforms that prevented
economic collapse. Nevertheless, thirty-one banks went under in
the space of two months. The banking reforms consisted mainly
of eliminating all the small banks and putting the sector entirely in
the hands of five large corporates—Mitsubishi, Mitsui, Sumitomo,
Yasuda and Daiichi. No sooner had this problem been addressed, than
the contagion of the Wall Street crash spread to Japan. Between 1929
and 1931, as Professor Bai Gao, a sociologist at Duke University has
explained, Japan's gross national product fell by eighteen per cent,
exports by forty-seven per cent, and investment in plant and equipment by thirty-one per cent. In 1930 alone, there were 2,289 labour
disputes and 906 strikes. By 1931, the net production of agriculture,

forestry and fishing was equal to only fifty-seven per cent of the
production levels just two years previously. As Bai Gao writes, 'many
families felt compelled to sell their daughters to serve as prostitutes.
In one hard-pressed village, one-fourth of the number of young girls
was sold; the village office was even used as a marketplace for transactions in prostitution'. He argues that the Japanese Army provoked
the 'Manchuria Incident' of 18 September 1931—the explosion that
destroyed a section of a Japanese-owned railway line, thought to
have been planted by mid-level Japanese officers but blamed by the
Japanese on Chinese nationalists as a pretext for invasion—with the
intention of (the army) surviving the economic crisis at home by
expanding Japan's military power in mainland Asia.

The economy had also been weakened by trade deficits that
persisted through the 1920s. Then the Hamaguchi government made
a dreadful decision: as of January 1930, the country was placed back
on the gold standard. Japan was faced with twin shocks—the spread
around the world of the Great Depression and the sudden appre-
ciation of the now gold-backed yen. Two months after the gold
standard had been reintroduced, *The Economist* was able to report
the consequences: in the first thirty days after the gold standard was
reintroduced, the Bank of Japan saw a huge outflow of the yellow
metal from its reserves and, because paper money had to be backed by
those reserves, a sudden contraction in the amount of money in circulation, the very opposite of what was needed to keep the economy
on an even keel. The magazine noted that 'the numbers out of work
have risen sharply and the move for wage reductions has seriously set
in. Trade disputes have naturally arisen in great numbers'.

Panicked, the government in December 1931 brought back as
finance minister the former prime minister Korekiyo Takahashi.
He ended the gold standard on his first day back in power.

But, on that same day, Japan had already been at war for almost
three months following the 'Manchuria Incident', the Japanese
Kwantung Army quickly subduing large parts of Manchuria.

Meanwhile, Takahashi managed to revive the Japanese economy.
The yen was allowed to depreciate by sixty per cent against the

United States dollar and forty-four per cent against the pound. He had the Bank of Japan underwrite government bonds to push more liquidity into the economy and then cut interest rates in successive moves. Then, after the economy began growing again, Takahashi decided to reduce government spending in order to avoid the ravages of inflation. That included curbing military budgets. The end result was that on 26 February 1936, a group of ultra-nationalist army officers assassinated the 81-year-old Takahashi in front of his wife, shooting him seven times and slashing him with a sword.

His successor, Baba Eiichi, did not make the same mistake. He doubled the military budget and so effectively cleared the roadblocks to war. The effect on the economy was devastating: for the next few years, inflation ran at about fifteen per cent a year but after 1941 hit an annual rate of 57.8 per cent. By 1938, seventy per cent of the Japanese budget was committed to military spending. It was a burden that could not be borne for long.

Fiscal management aside, there was another serious flaw in economic policy that left Japanese industry handicapped as it worked to supply the ever-growing war machine. That problem was tariffs on imports. By 1933, importing ships attracted a tariff of 37.5 per cent, and there was a 24.1 per cent impost on iron and steel. It was the tariffs on imported minerals that so exacerbated the shortages later on.

As a Japanese economist, Osaka Gakuin University Professor Yasukichi Yasuba, was to reflect in 1996, 'there is little doubt that the tariff on pig iron helped bring disaster to Japan'. And not just from the point of view of supply: Yasuba argues that the drive for self-sufficiency in the production of pig iron later encouraged the military to launch the Pacific war. Earlier, of course, the military's tune was quite different: it justified imperial expansion on the logic on Japan having insufficient resources. And, naturally, the military build-up just heightened the demand for those resources.

Foreign loans were an issue, too. As *Time* magazine put it in April 1939 Japan 'has not been able to wring a single yen from her busy but broke allies, Germany and Italy'. But China, even after it had lost Shanghai and much of the north, was still able to tap other countries for loans. In late March, Britain had loaned Nationalist China £10 million; the United States had provided $25 million and Belgium £20 million. *The Times'* correspondent in Shanghai was reporting at beginning of April 1939 the establishment of a non-official barter agreement with Nazi Germany; under this, China was intending to export raw materials in return for manufactured goods and railway equipment.

Extraordinarily, the Japanese government did not tackle industrial organisation until 1940. Throughout the 1930s, corporate Japan was chaotic. Employees were prone to seek better conditions by switching jobs, business bankruptcies were common, banks sometimes failed, and the regulatory system was weak. It was only in 1940 that Tokyo brought in regulations to force people to stay in their present jobs, introduced seniority pay to underline the notion of lifetime engagement, and reduced the power of shareholders to call the shots.

## IV

Before it attacked the United States, and then the Asian colonial territories of the European powers, Japan had years to put in place an ordered, systematic war economy. Yes, there were plenty of hands, willing and unwilling. Apart from the 100 million Japanese, Korea was another source of not only soldiers but civilian workers. As Paul H Kratoska writes, 'millions of people in northern China worked for the Japanese in agriculture and construction, with some being sent to Manchukuo, Mongolia and Tibet. There was also widespread recruitment of civilian labour in Taiwan and the occupied territories of Southeast Asia'.

It was astonishing that, just five months before the Japanese aircraft struck the U.S. fleet in Hawaii and brought America into the war, Japan was finally (and frantically) buying raw materials—but not necessarily for itself but rather serving as a conduit to Nazi Germany. Around 4,000 tons a month of copper and some cobalt was being shipped from Chile in Japanese ships and being delivered to Vladivostok for railing to the Germans. By contrast, Japan in 1939 had bought just 150 tons of copper from Chile. It was believed by the Allies that tungsten and manganese purchased by Japanese agents may also have been intended for their Nazi allies.

One sector did, however, expand in the pre-war years and kept on doing so up until 1941: heavy industry. This sector had gone from constituting thirty-eight per cent of Japanese manufacturing in 1930 to seventy-three per cent 1942.

Yet, here again, the lack of organisation for a war-footing on the home front was frighteningly apparent. N. Skene Smith, formerly a professor of economics at the Tokyo University of Commerce, wrote in *The Financial Times* in 1938 about the general inadequacy of the industrial workforce. He explained that some fifty per cent of all workers in Japan were engaged in fishing or farming, and another thirty per cent were in the professions, the civil service or in transport. That left just twenty per cent of the workforce in industry; but that has to be qualified by noting that about half of those six million people in industry were engaged in small firms providing food, simple housing, furniture and clothing.

There was also sabotage in industry, not directly intended to wreck the Japanese war effort but rather to profit from it. Key materials were diverted away from war production. Manufacturers obtained raw materials as a priority right, but there was padding of orders beyond what was needed to produce the war materiel required, thus allowing companies to divert what was left over into producing civilian goods to be sold on the black market. Japanese historians have estimated that in 1940 as much as a quarter of steel supplied was diverted to civilian production. The war effort was further hampered by individual ministries jealously guarding their own authority and resisting centralised control.

As Richard Rice, then assistant professor for history at the University of Tennessee wrote in 1979:

Compared to other nations, wartime Japan did poorly in establishing a single administrative structure capable of centralising

economic decisions and thereby directing the entire economy
toward production goals established by the military. The underlying
cause was lack of a popular political consensus. In Germany and
Russia, totalitarian governments centralised decision-making
power—although not until 1942 in Germany. In the United
States and Great Britain, wide and enthusiastic support for war
aims permitted suspension of peacetime constitutional freedoms
in order to create relatively efficient administrative structures for
economic mobilisation.

While Japanese corporates fought to defend their own patches and
interests, by contrast there was in the United States a pooling of
business expertise. Roosevelt first pulled in William S. Knudsen,
president of General Motors, to join the National Defence Advisory
Commission. Other members included Edward Stettinius Jr., president of General Steel and representatives of the Federal Reserve,
Securities and Exchange Commission, the Chicago Burlington
Railroad, the dean of women at the University of South Carolina
and the president of the Amalgamated Clothing Workers.

As Arthur Herman has chronicled in his study of American
industry response to the war effort, in spite of the Great Depression
the U.S. had a reasonable head start. In 1939 its factories turned
out three million cars a year, its resources sector 'produced more
steel, aluminium, oil and cars than all the world's great powers put
together'. But, of course, more—much more—would be required.
After all, Germany had been preparing since 1935, Britain from 1936.
Herman adds that Japan had been preparing for long before even
that; but, as we have seen, not to all that great effect.

## V

The military planners in Tokyo had assumed they would need
enough raw materials for two years of fighting, after which they
would have conquered the countries which produced them. It was a
miscalculation compounded by other factors. Unlike the Germans,

the Japanese never put their heart into synthetic oil manufacture and this peaked at just one million barrels in 1943. It was not a matter of lack of coal on the home islands: most of the synthetic production was based in Manchuria, where there was coal.

So many oil tankers were sunk that imports into Japan of fuel virtually ceased by January 1945. The Japanese turned to pine root oil, but none of this ever reached a refinery. As the Americans attacked Okinawa, there was oil sufficient to get only one (the *Yamato*) of the five still floating battleships to sea. Half of the oil-burning merchant vessels were laid up by the beginning of 1945, exacerbating the supply situation; deep sea fishing was abandoned.

As Jerome B. Cohen explained just after the war in the *Far Eastern Journal*, how different it had all been in 1941. In 1930, the Japanese motor industry had turned out just 500 units (cars, lorries and buses); in 1941, they managed 48,000 vehicles. In the same period, annual aircraft production rose from 400 a year to 5,000 aircraft. An aluminium industry had been created from scratch. There were similarly impressive achievements in the steel, power generating, oil refining and both naval and merchant shipbuilding. In 1940, the production of organic high explosives was greater than that in the United States.

Conquering North China meant ready availability of coking coal and by 1941 the occupied areas of China were supplying forty-one per cent of Japan's iron ore. Pig iron production in Manchuria was more than tripled. Soybeans from Manchuria became an important food source for the home islands, and almost all the salt consumed in Japan came from northern China. All these achievements, however, were impressive but came nowhere near what the British and Americans were able to do with their backs to the wall.

In the race for resources as war approached, the Japanese had the help and co-operation of a British-owned company, the giant Kailan Mines Administration. Its coal mines, inland from Tsientsin, were to prove vital to Japanese energy needs. Coal exports from North China to Japan rose steadily from 550,000 tons in 1935 to 2.4 million tonnes in 1939 and the British company was aiming at

selling Japan more than four million tonnes in 1940. The seventy-
two British employees of Kailan Mining, including Chief Manager
Edward Nathan OBE, remained at their posts and were allowed to
keep running the company and its 47,000 Chinese labourers, producing coal for the Japanese war effort and training Japanese in mine
management. When Nathan was replaced in 1943, he received a
glowing letter from the Japanese legation in Peking and was interred
in Shanghai for the remainder of the war. Incidentally, even after
Japan attacked Hong Kong and other British territory, the British
staff of the Hankow Power & Light Company continued to operate
that utility under Japanese control.

The growing Japanese influence in the north led to supply disruptions to the rest of China as more coal was shipped across the Sea
of Japan. Previously, most of Kailan's output was transported to Hong
Kong and Shanghai but now the Japan Iron Manufacturing Company took precedence. The British company made very substantial
profits from Japanese sales up to the end of 1941.

Hainan Island, off the southern coast of China, also had its part
to play in supplying Japan with iron ore. Between 1941 and 1944,
the Japanese government provided almost sixty million yen to build
276 km of railway lines in southern Hainan linking the iron ore
mines with the port of Sanya. After the construction of sections
with military significance were completed, Japanese troops killed the
press-ganged Chinese labourers involved. Work began on exploiting
the resource in 1939, and by 1943 it was possible to produce close to
one million tonnes a year of iron ore. However that peak was reached
only that year; in 1944, production fell to 352,436 tons. It was a
short-term proposition: by the end of the war the resource had been
largely worked out, and twenty-six locomotives along with around
one thousand rail wagons were idled. Japanese mining companies also
engaged in producing copper, tin, silica and limestone; the first of two
planned cement plants was completed in December 1944.

Formosa (now Taiwan) had been a Japanese colony since 1895,
and was a vital link in the occupier's economic chain. It was seen as
Japan's rice bowl, and the policy from Tokyo was to use the island to

capture a large part of the world sugar trade. The island became the world's fourth largest exporter of sugar.

In 1944, the population of the island stood at 6.7 million. Of the 450,000 Japanese who lived on Formosa, all but 2,000 technical and management people had been repatriated to the home islands. Japanese rule had seen considerable industrialisation: when the Chinese took over in 1945, they acquired the formerly Japanese built and owned sugar refineries, a pulp and paper industry (using the 200,000 tons a year of bagasse from the sugar plantations), cement plants and fertiliser manufacturing. The water level of Sun Moon Lake had been lifted by thirty-seven metres to fill a master dam, and total generating capacity in Formosa at the end of the war was 321 megawatts. Many of the forty-five sugar factories had been converted to producing alcohol for fuel. The pulp and industry's output was equivalent to about one-third of the entire output from the Chinese mainland. Three cement plants were in operation, the newest having been opened in 1943.

Back on the home islands, the government had stockpiled what seemed to be an impressive range of commodities: this included fifty-one million barrels of oil, 2.6 million tonnes of iron ore, 254,740 tons of bauxite, 5.7 million tonnes of scrap iron and steel and 81,000 motor vehicles.

But this did nothing more than paper over Japan's vulnerability. Getting control of the raw materials was one thing, doing something with them quite another. While Japan's seizure of large swathes of mainland Asia, the Philippines and the Netherlands Indies gave it access to oil, coking coal, bauxite, rubber, nickel, tin, cotton, phosphate, potash, magnetite, graphite, food and ferro-alloys, time and capital was going to be needed to exploit these resources fully. The problem that faced the Japanese was that few plants existed in the occupied lands with which to process the raw materials. It has been argued that the conquered areas rich in resources were important to the war effort for only the first eighteen months of the war phase that began in December 1941, the shipping losses eventually making it increasingly impossible to move the required resources. Once those shipping losses mounted, the Japanese economy became increasingly dependent on what could be supplied from northern China and from within the home islands.

By 1944, the military leaders in the various theatres learned they would get no more heavy artillery, that ammunition could no longer be used in training, and that they would receive severely reduced supplied of medium tanks and armoured cars (and no more light tanks).

It appears there was simply too much complacency in Tokyo, an assumption that conquest would solve the problem of raw materials. As early as 1941 there had been an actual reduction in the number of keels laid for new merchant ships. Figures compiled by the Japanese Iron and Steel Control Association told the story: in 1937, finished production of steel within Japan was 5,147,000 tonnes; yet in 1941 and 1942 the output was about the same, and rose slightly in 1943 to 5,609,000 tonnes. The impact of shipping problems and the general disruption caused by the fighting was shown by the flow of iron ore from Malaya and the Philippines; these countries had—when they were still volunteer exporters under the control of their respective colonial powers—exported three million tonnes in 1940 but in 1942, when controlled by Japan, they managed only 150,000 tonnes. By 1943, ninety per cent of the iron ore arriving in Japan came from China. These tonnages started to fall precipitously in the middle of 1943, and by December 1944 just 37,000 tons crossed the East China Sea. The steel problem was exacerbated by the coal problem, with coking coal shipments from northern China falling sharply as 1944 progressed, which not only reduced the quantities of steel produced but the quality. By January 1945 coal shipments from China had ceased.

The complacency leached into several sectors of the war effort. Apart from the failure to build up the production of synthetic oil products, as the Germans had done, the Japanese—again, unlike the Nazis—made little or no effort to repair factories damaged by air raids.

## VI

If you are going to invade another country, then probably one lesson from the Second World War is to choose a victim that can pull its weight, not one that you have to rebuild. It makes life so much easier. Czechoslovakia had all that heavy industry (including munitions plants), France all those paved roads and extensive rail network (and rolling stock that could be added to the Reichsbahn, the German railway system, all of western European operating on the same rail gauge), Denmark all those prosperous farms capable of helping feed Germany. Contrast that the invasion of the Soviet Union: yes, the conquered territories yielded much from oil to manganese to food, but the Russian railways used a different gauge and the distances for haulage back to Germany were immense, especially on an already stretched railway system. And then there were the roads: those same ones that feature in all of those photographs showing German equipment bogged down in quagmires as the unpaved roads turned to slush.

Japan took on a similar challenge with China. There were railways, but not enough; most Chinese still relied on river boats, pack animals and human carriers for their transportation needs. In fact, had there been better railways and highways in the remote provinces of Yunnan, Kweichow (now Guizhou), Szechwan, Shensi (now Shaanxi) and Kansu (now Gansu), the Chiang Kai-shek government and its armies may have been overwhelmed in their western China redoubt.

Northern China, including what became the Japanese puppet state of Manchukuo, was reasonably well-endowed with railways, at least in terms of other parts of Asia. But out west, and in the south, the network was minimal and six inland provinces had no railway lines at all.

## VII

It was one of those lines that no one heard before it was said in a movie, but it has now acquired the patina of being an accurate historical quote—it has the the attribute of something that should have been uttered at the time even if it was not. In *Tora, Tora, Tora*, the Japanese admiral, after his aircraft had returned from Pearl Harbor on 7 December 1941, and he has ordered his ships to turn for home, utters the words 'We have woken a sleeping giant and filled him with a terrible resolve'.

What the Japanese attack did trigger was the galvanising of America's industrial might.

The Falk Corporation, located in Milwaukee, Wisconsin, was a typical example. The then largest producer in the United States of gear drives had been through some tough years during the Great Depression but by 1938 it was starting to get back into the black. Then came Pearl Harbor. At least one worker, Les Greget, never had a day off for the next four years. The company was the beneficiary of the U.S. naval building programme that had begun in the late 1930s and went into overdrive in 1942. Falk was inundated with orders for marine gear drives. With $6 million in hand from the Navy, the company built an additional machine shop by early 1942 and worked it three shifts a day. Later that year, the female workforce had blown out five hundred. The company employed many African-Americans along one hundred and twenty-five Jamaicans brought to the U.S. as part of a government recruiting drive on that Caribbean island.

It was a story, in one way or another, replicated across the United States. In mid-1942, the director of operations at the War Manpower Commission, Brigadier-General Frank J. McSherry, said he was aiming at inducting close to twenty million people into war activities by the end of 1943. In 1942, war industries would need another 10.5 million people; on top of that, an estimated 3.4 million men called up for military service would need to be replaced.

Yet the Japanese destroyed the main centre of industrialisation in Asia outside their own country. Around seventy per cent of China's industry in 1937 was concentrated in the area between Wusih (now Wuxi) and Shanghai, the distance between the two being 125 km. The invaders closed most factories, leaving untouched only those owned by Japanese or British companies. Some 2.8 million Chinese factory workers were left without jobs—and the Japanese war machine without an industrial base in China. Destroying the industry was the means of subjugating China, and Japanese bombers concentrated on industrial plants. True, the loss of the industry crippled China's capacity to provide goods for its own people and the Nationalist military, but in the end it was self-defeating for the Japanese, too, when the destruction of merchant shipping limited the raw materials that could be transported from China and the finished goods brought back to China for the Japanese army. There were few factories left in occupied China that could take up the slack.

It was said that, until the Japanese army organised the Marco Polo bridge incident that gave them the flimsy pretext to invade China proper, no one in Tokyo had given any thought as to how China's economy could be harnessed for the Japanese war effort.

For the next few years, it showed.

# 3.

# The Race for Resources

BRITAIN IN 1939 WAS not quite as unprepared for war as post-war myth suggests. In March 1938, the Chamberlain government had published a white paper which set out an estimate for defence outlays of £343 millions, a record for British spending during any one year of peace, and of which the Royal Air Force took nearly £103 millions, the army £106.5 millions and the Royal Navy £133.5 millions (far more balanced, incidentally, than in the lead-up to the First World War where the Royal Navy received far more than the army, leaving the latter unprepared and poorly equipped for fighting in France). Since 1936, the number of RAF squadrons had increased from fifty-two to 123, and the number of people employed in the country's aircraft industry had tripled to 90,000.

Meanwhile, the Ministry of Supply had became an active trader in raw materials and by the beginning of 1940 had acquired about £150 million worth of commodities. Britain went into the war with large stocks of aluminium, zinc, antimony, silver, phosphate rock and other minerals. In fact, before the Asian colonies were overrun by the Japanese, the British Empire was claimed by *The Times* in early 1940 to be close to self-sufficiency in many materials. Between 1914 and 1939, empire production of copper had risen from 85,000 tons a year to 600,000 tons; with the metal lead the comparative figures were 130,000 and 630,000 tons, with nickel 25,000 and 95,000 tons, while aluminium had risen from 13,900 tons to 88,000 tons and asbestos from 98,000 to 344,000 tons. This meant the ministry could pretty well source most of its needs from empire producers; copper and zinc had been imported from Australia, Canada, Northern Rhodesia (now Zambia) and Burma, for example.

Wool (most of the Australian and New Zealand wool clips in 1938), flax, fertilisers and molasses were among other items that had been quietly bought up by the ministry (along with two million yards of linoleum, which showed some foresight: clearly someone had realised that new military buildings would need to have concrete rather than wooden floors—timber was a vital item—and so those floors would need to be covered).

The case of flax is instructive as to the depth of the economic potential of the British Empire. Linen flax was needed by the British for aircraft fabric, canvas goods, thread for boot manufacture, para-chutes and fire hoses. Russia had banned its export in 1938 and the the invasion of Poland and the Netherlands cut off supplies from those countries. The call went out to New Zealand where flax had long been produced but where the industry had collapsed during the Great Depression. The government in Wellington got behind efforts to re-open plants while farmers were encouraged to take one or two acres of their land to grow flax, for which a good price was paid. Flax operations sprung up over the country; there were seventeen plants built in the South Island alone, the machinery for pulling the flax being designed and built at the New Zealand Railways workshop at Christchurch.

Germany and Italy, on the other hand, faced one ineluctable factor: that Europe 'could provide no cotton, wool, hides, jute, rubber, coffee or tea', as *The Economist* put it. To go to war without them, or any prospect of acquiring regular and adequate supplies, was injudicious to say the least. That said, German scientists worked wonders. They developed a range of synthetic products that served in place of various metals, explosives, leather, rubber and, of course, oil.

*Women at work at a flax factory in Marlborough province in New Zealand.*
Marlborough Historical Society - Marlborough Museum Archives.

Even with the slowing production of coal, steel and aluminium, between 1942 and 1945 Germany managed to triple armaments production. And such was the standing of the leading German scientists that several thousand of them were employed by the victorious Allies after the war.

While all three Axis countries at various stages in the process of their defeat faced crises in supply of raw materials, none of the Allies suffered correspondingly catastrophic inadequacy of supply. Yet this was notwithstanding the fact that Britain imported most of her raw materials (with some exceptions, like coal). Even the United States was deficient across a wide range of raw materials (as it remains today, especially in regard to some strategic metals); as the war began Washington very much relied on foreign production of rubber, manganese, bauxite, chromium, copper, zinc, cobalt, tungsten, tin, antimony, sisal and jute. In the case of tin, the Americans depended almost entirely on supplies from British Malaya and the

Netherlands Indies (now Indonesia); domestically, the United States produced only 0.2 per cent of its tin requirements. By 1940 it was consuming 75,000 tons a year, or about half the world's supply and Washington was looking to Bolivia to help fill the gap. That year President Franklin D. Roosevelt asked Congress for $25 million to buy strategic metals. America obtained all but five per cent of its rubber from the Netherlands Indies.

The British Empire, the colonies of France and Belgium and Latin America (along with development of substitutes such as synthetic rubber) would make up those deficiencies. But getting them to Britain or North America was often a struggle in itself with the sometimes critical shortages of shipping space. Serious shortages of pulp and hides emerged in 1943, providing new worries for those charged with providing for the war industries.

## II

German and Japanese planners had a much taller order to tackle. The Nazis talked incessantly of 'lebensraum', in the context meaning 'living space'. The lands of the east were to provide German farmers with the vast acreages needed to supply the Fatherland with plentiful food and the man on the land with sufficient space. But the invasion of the Soviet Union also meant *autarkie*, spelled autarky in English, by which is meant independence from trade with others, a form of self-sufficiency—in this instance, all the necessary raw materials needed to supply the German war machine from oil for aircraft, tanks and U-boats to iron ore and other metals critical to the manufacture of those fighting machines, along with every gun supplied to the infantry and every bullet or artillery shell fired, along with all the food needed by the populace. It did not quite get near the Utopian ideal implied by the word: Germany never shook itself of dependence on the neutrals in the cases of iron ore from Sweden, chrome from Turkey, or tungsten from Portugal.

In July 1937, *The Economist* reported on German efforts to ramp up mine production. The magazine pointed out that, as a result of territorial losses after 1918, the country had surrendered to Poland three-fifths of Germany's Silesian lead and zinc deposits and four-fifths of her Silesian coal reserves. Germany had been the dominating producer of the fertiliser material potash but her mines in Alsace were now part of France (and so ending the monopoly Germany had on potash supplies), as were the previously German iron ore mines in Lorraine (which increased Germany's interwar reliance on iron ore imports from about sixty per cent of its needs to close to eighty per cent). An attempt was made to partially replace this with using very low grade ore at the Hermann Göring Iron Works, but this was clearly uneconomic.

Mind you, Germany by 1937 still mined 65.5 per cent of the world's potash, twenty-two per cent of the world's coal, along with appreciable quantities of iron ore, copper, lead, zinc and graphite. Scrap steel recycling was a significant contributor to the armaments drive—but also to manufacturing of the array of machinery needed if Germany was to be able to produce such key items as synthetic rubber, artificial silk and plastics.

Not that Germany was without resources. German mines produced 910,000 tons of iron ore and 15,600 tons of copper. But the Reich imported all its manganese, which meant capturing the Ukrainian mines was a critical goal (and that supplied enough to meet all German needs even though the metal had to be hauled all of 2,400km by rail back to Germany). Even by 1943, Greater Germany was providing twenty-three million of tons of coal that year against 6.4 million supplied to the Reich from the occupied territories.

So Germany went into the war with a surplus capacity of coal and potash production, but was partly or entirely deficient in every other material, particularly iron ore and petroleum. A study in 1939 from the Royal Institute of International Affairs showed Germany was producing about twenty-five per cent more coal than it needed, this providing a valuable source of potential export income or a basis for barter with other European states. As for copper, Yugoslavia was the obvious answer but the mines there were controlled by the French, and the main lead and zinc mine by British interests.

The other problem was that Yugoslavia, while Europe's largest source of the red metal, could fulfil only a part of Germany's copper needs; for the balance, this had to come from across the oceans, the world's main sources of copper then being the United States or Chile, along with Northern Rhodesia (now Zambia), Canada and the Belgian Congo. After September 1939, none of those sources would be available to German industry. Tin was a similar challenge: it needed to come from Malaya, Netherlands Indies (now Indonesia), Bolivia, China, Nigeria or the Belgian Congo. Vanadium was then mined only in South-West Africa (the colony that Germany lost after 1918, now Namibia), Northern Rhodesia, Peru and the United States. The metal lead would not be a problem provided Germany could get keep buying from a mine in Yugoslavia. In 1926, a British company, Selection Trust, headed by an American-born mining engineer by the name of A. Chester Beatty, gained access to Yugoslavia to an area at Trepca in Serbia which had been the site of historic mining. The Stantrg mine was opened in 1930. A separate company, Trepca Mines, was floated in 1929 with a capital of one million pounds.

When the shareholders of Trepca Mines met for the annual meeting on 29 December 1939 at the Chartered Insurance Institute, located on Aldermanbury near the Guildhall in London, Beatty was very positive about the mine. A new area, 435 metres below ground, had been opened up with high grades of lead and zinc, along with copper and silver. The first stage of a lead smelter had just been brought into production.

But the Yugoslav government, now concerned about its own security, needed money to buy modern weapons, and Germany was offering to sell these to the Yugoslavs. Trepca Mines had to agree that part of its production would be sold to the Yugoslavs for payment in dinars, the rest of the output being sold abroad as usual and the proceeds remitted to London. One of the main customers for the metal concentrates was Germany.

Trepca was an important source of lead and zinc; the country also had the Bor copper mine owned by French interests. In 1939, Bor was the largest copper producer in Europe; in the first ten months of 1939 its output reached 50,000 tons of copper. In 1938, 8,000 tonnes of its copper was bought by Germany but in October 1939 a German trade mission wanted the the quantity of minerals sold by Yugoslavia to its industries to be doubled, offering the Yugoslavs modern artillery and Messerschmitt aeroplanes in return. It was reported by *The Washington Post in* November 1939 that Yugoslavia was desperately trying to get armaments: they had hopes of buying second-hand tanks from Spain, and the British were being pressed to supply engines for the assembly of Blenheim bombers in Yugoslavia, while Belgrade was looking to Italy to supply Caproni light bombers from the Savoia company.

The expansion of zinc smelting in the years leading up to the war made Germany close to self-sufficient. The invasions of Czechoslovakia and Yugoslavia would ensure supply of antimony, the metal with the important quality of being a flame retardant.

Germany had another bounteous source of supply: the Union of Soviet Socialist Republics. Berlin under the 1939 non-aggression pact with Stalin had access to the vast food and mineral resources of the U.S.S.R, even though the Russians never admitted to the vast amount of aid they were giving the Germans; as late as just a few months before the Wehrmacht started its invasion of Russia, Izvestia was proclaiming that the Soviet Union was selling Germans nothing but grain. In fact, the Reich was receiving oil, iron ore, scrap iron, chromium, phosphates, food, timber, cotton, manganese and much else. But the Soviet leadership stuck to their story that only humanitarian goods were involved and they rejected the requests from London and Paris to join the blockade of Germany. There was a certain irony in Moscow's explanation that it was motivated by humanitarian concerns, but stick to that story they did, as George Ginsburgs, then of the State University of Iowa, explained in a 1962 essay on the Soviets and international law:

> Instead of admitting that their export to Germany consisted of strategically important raw materials, such as oil, iron ore, scrap iron, chromium, manganese, platinum, phosphates, foodstuffs,

cotton and lumber, and claiming that this was in no way contrary
to traditional international law, except that under the same code
Paris and London would have the right to intercept and seize
those cargoes as absolute or conditional contraband, the Soviets
sought instead to represent their commerce as dealing wholly in
goods destined for mass consumption by the civilian populace.

Not only were the resources of the Soviet Union being sold to the
Nazis, but the Soviets allowed goods for Germany to be railed from
the Pacific coastline via the Trans-Siberian Railway. These goods being
shipped across the Pacific to Vladivostok were the only part of the
German supply chain from the east which the British could attempt
to disrupt. In January 1940, Royal Navy ships stopped the Russian
steamer *Selenga* of 2,492 tonnes which was laden with antimony, tin
and tungsten sailing from Manila and bound for Vladivostok. She was
intercepted by three British cruisers and forced to divert to Hong
Kong. This was quickly followed by the stopping of another Russian
ship, the *Vladimir Maiakovski*, laden with copper and molybdenum
from Mexico. Subsequently, the Soviet ambassador in London, Ivan
Maisky, called on the Foreign Secretary, Lord Halifax, to protest the
seizures. The British initially refused to budge, saying they believed
the cargoes were bound for Germany. However, after a seventy-
five day internment, the British gave way in the face of continuing
Russian protests and the *Selenga* was allowed to continue her journey.

And the Russians blithely kept on fuelling the German war
machine, oblivious to what was to come. Nationalist China's Chiang
Kai-shek passed on reports he had obtained just days before Barbarossa
was unleashed by Hitler, but Stalin dismissed them, suspicious that
Chiang was trying to goad him to attacking the Japanese.

## II

Soon, of course, the Germans would be helping themselves—and
the invasion of the Soviet Union deprived not only the Soviets of
the raw materials in the areas overrun by German army groups, but

others who depended on it, too. The Americans produced only a small fraction of the manganese (for which there is no substitute in the steel-making process) consumed by their industries; in 1934, this output from U.S. mines had been as little as 853 tons. Most of America's deposits were low grade manganese dioxide, hard to process. As late as 1940, the U.S. had obtained about a quarter of all its manganese from the Soviet Union. Even if the mines had not been overrun, the German invasion had closed off the ports on its Black Sea coast even before the German armies captured the Nikopol manganese deposits in the Ukraine. Cuba became a critical supplier; the other sources, including the Gold Coast (now Ghana), South Africa, Brazil and India, all involved long-distance transport through seas infested with U-boats. Federal authorities were reported at the time to have stockpiled about two million tons of manganese, which was expected to be sufficient for two years' consumption. But no one knew, one, how long the war would last and, two, what might be the growth trajectory of the steel sector. Clearly new long-term sources needed to be found.

Control of the Nikopol mines increased Germany's manganese supplies by a third. Romania, as is well known, was an important source of oil for the Nazi regime; but the country was also a significant producer of manganese, bauxite, lead, copper and molybdenum. Yugoslavia, as we have also seen, was an important metals source, producing as well as base metals bauxite, some manganese and chrome. Greece brought with it enough nickel production to meet a third of the Reich's needs, while Hungary had bauxite, and the alliance with Finland would offer Germany access to that country's nickel and, more importantly, wood pulp.

But grabbing industries brought with it further problems. Belgium had well regarded metallurgical industries. Iron ore for those Belgian factories came from Luxembourg and Lorraine, also invaded by the Germans in May 1940. However, those steel mills also needed supplies from Sweden, Norway and Algeria. Belgian smelters sourced their zinc from Romania, India, Spain and Sweden. Copper and tin came from the Belgian Congo, a source cut off by the British blockade. In fact, Belgium produced very little of the raw materials on which its metals industry was built; only cement, brick and glass was supplied from sources within its own borders. The Netherlands, too, had no deposits of iron, copper or other ores and its metal industries depended on imports; a large range of manufactures had been developed using tin from the Netherlands Indies, tin that was no longer available after the German invasion. Holland also lost a quarter of its export business—the goods that had formerly been sold to Britain. So the metallurgical industries of Belgium and the Netherlands could not be used to supply enough of the critical supplies for the war effort for lack of raw materials. For Germany, the war was just one big economic knock-on effect.

### III

One of the urgent issues for the Germans was oil. It would be in short supply until the Romanian fields could be controlled by Berlin. The invasion of the Soviet Union would mean vastly greater supplies would be needed; in fact, it would be the desperate lunge for the Caucasian oil fields that would place immense strains on the German army's fuel management.

Going to war meant needing to get more oil. While that sounds an obvious truism, and much would hinge on access to Romanian oil fields, the coming conflict meant a radical overhaul of the German oil industry. Oil had been produced from fields in Germany since the middle of the nineteenth century, but in 1938 these domestic fields were the source of just three per cent of the country's energy; ninety per cent of energy produced in Germany that year came from coal. In the 1930s, about seventy per cent of oil used in Germany was imported. By 1944, thanks largely to the growth of the manufacturing from coal of synthetic gasoline, Germany was 72.3 per cent self-sufficient, with sixty per cent of its oil supplies being supplied from the two synthetic technologies. The loss of Romania's oil as Soviet armies advanced, and the earlier failure to capture Russia's fields, were serious blows. Then in 1944 Allied bombers targeted the

synthetic plants inside Germany. It was not only the plants that were
bombed, but the dislocation of the German rail system resulting from
Allied aerial bombardment meant that not enough coal was reaching
the synthetic plants and much of what they produced could not be
distributed throughout the Reich.

The National Socialists did make some efforts to improve oil
supplies right from their early days in power. Exploration drilling
tripled between 1933 and 1937, although no significant discoveries were made. In 1932, Germany produced 230,000 tonnes of
oil (1.72 million barrels); in 1940, production of crude in the old
Reich was about one million tonnes (7.5 million barrels), this result
achieved largely through forcing higher rates of production from
existing fields. Following Anschluss, the Austrian refineries had been
forcibly merged with their German counterparts.

As soon as the Nazis came to power, they worked closely with
I. G. Farben to increase synthetic oil production and the new regime
provided support by raising fuel prices. Hermann Göring's Four-
Year Plan Office ordered the establishment of the Reich Institute
for Petroleum Research, located in Hannover. Eventually Göring
came to favour synthetic production as a means of solving German's
domestic shortage of fuel, mainly because he wanted to keep his
Luftwaffe supplied with aviation fuel. There was also the important
consideration from the start that producing synthetic fuel might
be expensive, but it at least saved Germany the enormous foreign
exchange cost of importing fuel. By 1939 the situation became more
urgent because, while Germany did get oil from Romania, the Nazi
regime (like Italy) depended mainly on supply from countries that
would soon be on the other side of the British naval blockade: much
Nazi oil by that time was still coming from the United States (this
was after the huge oil finds in Texas), Venezuela, the Netherlands
Indies, Persia and Mexico. After the Anglo-French blockade was
imposed in 1939, Germany relied on Romania and, under the terms
of the non-aggression pact with Moscow, the Soviet Union for oil.

There were two types of synthetic production. The Bergius process involved very expensive hydrogenation, where hydrogen was

added to coal under high temperature and high pressure. Then there was the Fischer-Tropsch technology where coal molecules were broken down into hydrogen and carbon dioxide. It was the Bergius process that produced the higher grade aviation fuel.

In 1936 Hitler announced ten new hydrogenation plants. By 1939, the production of synthetic oil reached the level of supplying about a third of the nation's oil needs; but, of course, this was still peacetime and the advent of war would send the consumption figures soaring. The 1939 study by the Royal Institute of International Affairs noted that Romania's production was declining (it had been doing so since 1936) and the other obvious source of supply, as seen from London, was from Baku in the Soviet Union—a conclusion that was also eventually reached in Berlin.

I.G. Farben had opened Germany's first synthetic gasoline plant in 1927. Its Leuna plant could produce 90,000 tonnes (675,000 barrels) a year. The business struggled, but was effectively rescued by the Nazis imposing higher tariffs on imported crude and thus making the Farben Bergius process economic; the advent of the Great Depression two years after the Leuna plant opened, along with big discoveries in Oklahoma and Texas having sent oil prices tumbling, had raised a large question mark over Leuna's survival.

As at 1937, Germany's production of petroleum and related fuels was still far short of its needs. The coal-oil plants, both those operating and under construction, provided a capacity of 940,000 tons a year; in 1936, the country's imports of oil and oil products totalled 3.84 million tons. Even with its initial conquests, in the case of part of Poland's Galician oil fields (the Russians had the lion's share of that field in their part of newly dismembered Poland) and those of Alsace (after France was subdued) increased the Reich's output only marginally. However, the subjugation of France delivered into German hands 250,000 tons of stored aviation fuel.

It was given as evidence in the Nuremberg trials that Hitler told his generals in early 1942 that, unless he could get control of the Caucasian oil in the south of the Soviet Union, then Germany would fail on the Eastern Front. The Wehrmacht was repulsed before it could take Baku and, although it captured the Maikop oil field (near the Black Sea), this had been destroyed by Russian troops, including having concrete poured down all the wells. By the time the Red Army recaptured Maikop in late 1941 the Germans had managed to extracted a mere 1,000 tons of oil from the field.

As Joel Hayward argued in his 1995 study of Hitler's quest for oil it was far beyond Germany's capability to provide enough fuel for the invasion effort in the Soviet Union. The supply task was enormous: it meant keeping on the move 3.6 million soldiers, 600,000 vehicles, 3,600 tanks, and more than 2,700 aircraft. Blitzkrieg, as in the conquests in Western Europe, used relatively little fuel because of the short distances involved and the short duration of the campaigns. The great expanses of Russia and Ukraine were quite another matter.

For the Japanese war machine, the petroleum supplied from the captured territories of the Netherlands Indies, Borneo and Burma was vital. But, as the League of Nations noted in one of its regular economic surveys, the 8.8 million tonnes produced in those three territories in 1941, along with the 0.4 million tonnes produced in Japan and Formosa, constituted less than two per cent of the 498 million tons produced in 1941 in the world (excluding Continental Europe). 'The production of Continental Europe in that year may have amounted at the most to fourteen million tonnes, including the synthetic oil', the League noted in its 1945 summing up of the war effort.

But the point is that the bulk of the world's oil remained within the control or influence of the Allies. The Soviet Union produced thirty-three million tonnes in 1941 and some of that capacity was to be captured or wrecked by the invading German armies.

However, as with metals and food, transport was a constant problem. In fact, Latin American oil producers were forced to reduce their output due to shortages of tankers on the Atlantic runs. The Americans partly made up for this by lifting production of their domestic oil fields; in 1941 output had been 191 million tonnes but by 1944 American oil fields were pumping at an annual rate of 227 million tonnes, some twenty-five times the amount of oil available to Japan.

## IV

But even while war was raging in Europe, and the British could disrupt German trade around the world, Japan was still stocking up. Within weeks of the attack on Pearl Harbor, much of Southeast Asia came under Japanese attack and, within a short space of time, occupation. This invasion cut off more than sixty per cent of the tin supplies to the United States, sixty per cent of its hard fibre and the source of practically all the rubber hitherto imported into America.

It was not just a matter of ensuring increasing production in the colonies and other parts of the world beyond the control of the Axis powers. There had to be economising. Take tin: the Japanese invasion of Burma, Malaya and the Netherlands Indies suddenly removed large supplies of this metal for the Allies. Yet researchers and metallurgists had been working on this eventuality and had been able to come up with the means of reducing its use; half the tin produced then went into tinplate (which still today lines the interior of cans containing food and beverages) but tin can manufacturers either resorted to using black plate (the steel or iron before being converted into tinplate) or by using the electrolytic process to apply the tin more thinly than had previously been possible with the hot dipping process. As *The Manchester Guardian* reported at the end of 1942, tin consumption for tinplate in the United States, which had stood at 48,700 tons in 1941, was thus able to be reduced to 21,000 tons by 1942 and for 1943 the ambition was to get it down to 11,000 tons. This, combined with increased output from Nigeria, the Belgian Congo and Bolivia, helped close the gap. By the end of 1942, also, it was reported that discoveries had identified tin deposits in Canada, Alaska and Mozambique, although nothing much seems to have come of these. All this should be seen against the background that supplying military forces in many theatres involved vastly increased production of cans for food.

For all of Japan's omissions in planning for its wartime commodity needs, there had been at least sporadic efforts to acquire minerals. In early 1935 a delegation from Nippon Industrial Company was in Western Australia. They sailed north from Fremantle aboard the

government motor ship *Kangaroo* to inspect iron ore deposits on
Koolan Island. Singapore's *Straits Times* reported the Japanese were
looking to Australia as an alternative iron ore supply to those it controlled in Malaya. Nippon owned the Dungan mine in Trengganu
which was exporting 500,000 tonnes a year while two other mines
were controlled by Ishihara Sangyo Koshi. The Japanese did make
progress in Western Australia but the business came to an end in
1938 when the Australian federal government banned exports on
strategic grounds. The previous year had seen trades unions protest
over Nippon's plan to replace local miners with Japanese workers at
Koolan Island.

## V

One statistic tells the story: the United States during the war needed
to import minerals from eleven countries in Latin America and fourteen territories in Africa. The importance of Africa in this conflict
is adduced by some other statistics; during the peak years of the war,
Africa supplied fifty per cent of the world's gold, nineteen per cent
of its manganese, thirty-nine percent of its chromite, about twenty-
four per cent all the vanadium mined—all those three being vital
to steel making—as well as seventeen per cent of world production
of copper, almost ninety per cent of cobalt, ninety-eight per cent of
industrial diamonds (mainly Sierra Leone) and, of significance in the
later years of the war, all of the uranium. When the atomic bomb
became a reality, the uranium came from Africa (the Belgian Congo);
uranium deposits of what became the Shinkolobwe mine in Katanga
province were discovered in 1915. This mine, near the southern
Congolese town of Likasi, produced uranium for the first atomic
bombs. (The Shinkolobwe uranium mine ceased was closed in 1960,
when Belgium granted Congo independence. Belgian authorities
filled the main uranium shaft with concrete.)

In Africa, the vital territories for mineral supplies were South
Africa, the Belgian Congo, the two Rhodesias (Northern and Southern), along with Nigeria and the Gold Coast. Britain required not

only the staples of industry—copper, chrome, platinum, bauxite, zinc and manganese—but metals needed for new specialised applications such as antimony (fire retardant), beryllium, graphite, mica, talc and fluorspar. The supply situation was exacerbated, first, by the loss of supplies from Europe (especially Sweden) and, second, by Japanese over-running of sources of tin, bauxite and tungsten in Asia. It was primarily the British colonies in West Africa that were critical to supplying the Allies with the metals they needed; French West Africa was far less important in this role. The tin lost through the occupation by Japan of Malaya, Siam and the Netherlands Indies and the seizure of Chinese producing areas left only two other major producers: Nigeria and Bolivia. The Gold Coast became an important supplier of manganese, being the third largest producer in the world (just behind the U.S.S.R. and India).

## VI

American planners faced a problem. Most of their tungsten came from China, and they needed to turn to both obtaining more from South America or finding deposits at home. Without tungsten, the defence plants could not make armour or ammunition sufficiently tough and heat-resistant. The metal was used in railway rails, chisels, hacksaws, armour plate and armour-piercing bullets and shells; in tools the tungsten content could be as high as eighteen per cent. By 1942, imports had dropped to 4,500 tons from 16,157 tons in 1940. Washington feared that, as Japanese forces conquered more of China, access to the latter's tungsten output would be cut off. (Back then two-thirds of the world's tungsten was supplied out of China.)

By mid-1937 *The Wall Street Journal* reported that tungsten's price had risen ten-fold since 1933. It had been the fierce fighting around Shanghai in 1937 that had really started the price soaring. Within a few days, the *Journal* ran a more extensive analysis of what it saw as a looming tungsten crisis, with the U.S. at that stage consuming thirty-one per cent of global supply (and American mines able to supply only eight per cent of world output).

The great success story over the next few years was that of domestic production, which increased by forty-two per cent in a year by 1940, with California producing large amounts of tungsten and ten other states getting involved in the industry. By 1939, the *Los Angeles Times* was reporting that miners and prospectors in Nevada had laid aside their exploration for gold and silver and were instead switching to tungsten, chrome, quicksilver (mercury), molybdenum, manganese, lead and zinc.

A Chinese importing company, Wah Chang Trading, was in late 1939 given a government contract tungsten and was to deliver the first consignment of 427 tons sourced from Hong Kong, Burma and Indochina. In addition, Chile began in 1940 exporting tungsten to the U.S. for the first time, and Argentina and Bolivia both lifted shipment volumes. In addition, the Americans were able to substitute molybdenum for tungsten in some steel products.

By 1943 the British Association for the Advancement of Science compiled a report for the Allied governments. They estimated the Allies' steel-making capacity as twice that of the three Axis nations put together, with the United States having ten times the capacity of Japan. Of the entire world output of critical metals, the Allied powers—the United States, the British Empire and the U.S.S.R—controlled or had access to ninety-seven per cent of the world's supply of vanadium, ninety-six per cent of molybdenum, eighty-two per cent of chrome and of cobalt, seventy-seven per cent of manganese and of tungsten, and about ninety per cent of the world's copper.

## VII

Gold, which *Time* magazine in January 1942 dismissed as something 'nobody really needs', was being mined at peak rates and taking labour, equipment and shipping space needed elsewhere. The magazine harrumphed that the British Empire had five hundred thousand workers employed producing gold, while the United States had fifty-five thousand. As to the latter, President Roosevelt would soon see to that: in 1942 he ordered closed all gold mines in the country on the grounds that mining gold was not a war priority.

For Britain, though, gold was very much a priority. Britain needed gold with which to buy U.S. dollars; the Americans, on the other hand, needed copper, zinc and lead—but not gold. After all, Britain was paying its bills to Washington partly in the yellow metal so Roosevelt could afford the luxury of closing domestic mines (and also ban export of mining equipment to gold companies). Of the half million subjects of the empire engaged in gold mining, four-fifths of that number were in South Africa (and other thirty-eight thousand in Canada—something else that rankled with *Time*).

When war broke out in 1939, South Africa was the premier gold producer of Africa with output reaching its peak in 1941 at 408 tonnes. Other important gold producers were Southern Rhodesia, Gold Coast, Belgian Congo, and Tanganyika. Almost all of South Africa's wartime production of gold was shipped to Britain. It was needed to bolster reserves and offset trade deficits.

But, as with so many other aspects of the war effort (military, political and economic), there was friction between Great Britain and the United States. The former placed great emphasis on keeping the South African gold mines going to help pay London's bills; the latter tried to apply pressure in Johannesburg to divert manpower and spending to much needed production of strategic metals.

As was the case with many commodities where the Imperial government dominated output, the price of gold was fixed from 1940 until 1945; at the same time, the miners were suffering from the rising costs of all their imported inputs. Once Britain was no longer desperate for gold with which to pay its bills when Lend-Lease was extended, gold prospecting was banned in the East African colonies, and the mining companies were denied priority for machinery. Moreover, labour was scarce. The impact of Lend-Lease hit the gold industries in both Kenya and Tanganyika; only since the 1990s have they begun once again to grow significantly.

But, of course, the South African gold industry had its own problems. White skilled workers and supervisors were going off to war, many of the machinery works that serviced the gold miners' equipment were turned over to munitions or other war work and—a problem that would get worse as the decades wore on—mining operations had to go ever deeper (and more expensively) as the more easily won, shallow gold was worked out.

Overall, though, gold production fell as the war progressed; in 1944, only about two-thirds of the amount of gold was being produced as in 1940. Part of the reason, as has been noted, was that gold mines in the United States, Canada, Australia and British colonies in West Africa were under government direction to restrict production in order that manpower could be released for other war work.

However, the American pressure for more strategic metals did have some effect: South Africa lifted output of not only platinum (needed in such things as spark plugs) but also of chrome, manganese, vanadium and copper.

Then there were diamonds, needed in the manufacturing process for so many weapons and items of materiel of war. Diamond drill bits, for example, were needed to pierce alloys and, unlike their decorative counterparts, industrial diamonds thus wore out and there was a constant need for replacements. The dominant producers of diamonds were the Belgian Congo, the Gold Coast, Angola and Sierra Leone, as well as South Africa.

## VIII

By 1940, the Americans realised that their $900 million rubber industry could face its first threat in its 101-year history. The problem for the United States manufacturers was more than ninety per cent of their rubber came from territories controlled by European colonial powers; the defeat of the Netherlands by the German army in 1940 provoked alarm at the effect this might have on rubber supplies from the Netherlands Indies. The other main sources were British Malaya and Ceylon. Only a tiny quantity came from rubber plantations in Latin America: in total, Brazil, Costa Rica and Mexico (along with Liberia in Africa) supplied just six per cent of American needs.

Britain controlled half the world's production of rubber in 1939, with 375,000 tons produced in Malaya, 60,771 tons in Ceylon, 16,112 tons in India-Burma and 24,544 tonnes in Sarawak.

decided before the European war began to start a stockpile and by
the end of 1941, when America came into the war, U.S. rubber
inventories had reached 533,000 tons. Within two months the supplies from Southeast Asia had been cut off and there were fears the
other rubber producing colony, Ceylon, could also be overrun by
the Japanese. In 1941, too, the United States government let contracts, each of around $1.25 million, to United States Rubber, B.F.
Goodrich, Firestone Tire & Rubber and Goodyear Tire & Rubber.

During 1942, Washington tied up as much Latin American
production as it could through the Rubber Reserve Company.
Private contractors were also lined up—Goodyear in Costa Rica,
Continental Mexican Rubber Co in Mexico and Chicle Development in Honduras. Brazil was provided with a $100 million credit
to enable further rubber plantation development in the Amazon.
Ecuador was encouraged to tap more wild rubber. The Americans
had their eyes on the wild rubber potential in Panama—where
there were an estimated four million such trees—and Colombia; the
*Chicago Tribune* reported in April 1942 that a mission had gone to
Guatemala to see whether the former wild rubber business could be
revived, and a contract was let for purchases from Bolivian producers.

The U.S. also put great efforts into Haiti, which Los Angeles
newspaperman Gene Sherman visited in late 1943 and described the
'frantic project' to grow cryptostegia, a plant which produces high
grade latex. He reported back to the *Los Angeles Times* that the plant
species had been imported into Haiti in 1912 from Madagascar, but
for its lavender smell and appearance rather than its latex. By the
end of December 1943, he said, it was planned to have 75,000 acres
planted for a 1944 production target of 3,000 tons of latex, rising to
12,000 tons in 1945. The output would be infinitesimal by American
needs but 'good news at a time when rubber bands are curios'.

Liberia played its part, too. When Japan attacked the United
States, Liberia was producing about 10,000 tons a year of rubber,
almost all coming from plantations owned by Firestone. In 1926,
the company had leased 4,140 square kilometres of jungle. Firestone

picked Liberia because its soils and climate made it ideal for growing rubber. The first ship of latex arrived in the United States in
December 1940 aboard the steamship *West Lashaway* (which would
later be torpedoed by U-66 in 1942 while carrying 7,670 tons of
much needed tin, copper, cocoa beans and palm oil from the Belgian
Congo to the United States). By 1944 Firestone had lifted production to an annualised 20,000 tonnes but the company's tyre plants in
the United States were depending on synthetic rubber for sixty-five
per cent of their feedstock. While the rubber supply problem had
been licked, the production of sufficient numbers of tyres had not;
suddenly in 1944 the army needed a large number of tyres as it
was by then engaged in large land operations in France, Italy and
the Pacific islands. Some expert tyre builders were released from the
forces, and *The Wall Street Journal* was reporting that $100 million was
being invested to increase output.

Brazil's President Getúlio Vargas, once he had decided that the
Allies were more likely than the Axis to win the war, threw his energy
into what he dubbed the Battle for Rubber. The plan was to harness
the wild rubber growing in the Amazon jungle; the plan served a
dual purpose in allowing the government in Rio de Janeiro to extend
its influence into the Amazon region, Vargas having in 1940 been the
first Brazilian head of state to visit there. (Brazil later also sent 25,000
men to fight with the Americans in Italy, with 457 being killed.)

However, extraction of the latex from the wild rubber required
a work force. Enter the 'rubber soldiers', workers who either volunteered or more likely were dragooned into going to the Amazon to
tap the trees. About 55,000 men living on arid northeast subsistence
farms were moved to work on the rubber; most of them never to see
their homes or families again. When the United States terminated
the rubber arrangement with Brazil in 1947, the workers were simply
abandoned and left to their own devices in the jungle zone. It has
been estimated that around half those drafted died before the war
ended, many from disease (malaria, yellow fever, beriberi and hepatitis
being the main causes) or animal attacks from jaguars and alligators,
or from snake bites. Their experiences were largely forgotten until

around 2005 when stories broke to the foreign press about the efforts being made by the surviving 8,300 rubber soldiers to get a long needed pension increase.

The *International Herald Tribune* in 2006 reported the case of Alcidino dos Santos who, one morning in 1942, was on his way to the market, the nineteen-year-old having been sent by his mother to buy vegetables. A Brazilian army officer stopped him and told him he was being drafted to harvest rubber in the Amazon, and he had to go that day. The newspaper interviewed him in his simple home at Rio Branco, the capital of Acre state in the Amazon. He recalled starting work just after midnight and going into the forest in the dark to cut the trees, returning later in the day to collect the latex. They were never paid (the money owed was theoretically being held in trust for the workers) but you had to work or you received no food. At the end of the war, with no money, they had little choice but to stay where they were.

A BBC report in August 2010 highlighted the problem again. In interviews with the some of the surviving rubber soldiers, one told the reporter that 'those who tried to leave were given their pay and told they were free to go. But down the road hired guns were waiting to shoot them and take their money back to the boss'.

At the beginning of April 1942, U.S. Secretary of Commerce Jesse H. Jones announced that the Rubber Reserve Company and the Defence Plant Corporation—both affiliates of the Reconstruction Finance Corporation—had awarded contracts to increase production of synthetic rubber to 700,000 tonnes a year. Twenty-five large American companies were involved in the plan to build new rubber factories, and they included the Standard Oil companies of Indiana, New Jersey and Louisiana, along with Dow Chemical, Goodyear Tire and Rubber, Humble Oil Refining and Hycar Chemical.

Two years earlier, Jones—then Federal Loan Administrator, and acting under powers assigned by the passing of the Reconstruction Finance Bill just days before—had announced U.S. purchases of large quantities of rubber and tin, over and above normal commercial quantities bought by American companies. The International

Rubber Regulating Committee, under Sir John Hay, had agreed to release to the Americans an additional 150,000 tons of rubber, about fifteen per cent of the annual output from the Netherlands Indies. The corresponding tin cartel, the International Tin Committee, had agreed to selling an additional 75,000 tons of that metal. A committee of experts commissioned by Roosevelt had reported to him, listing the alternative supply sources for what these men regarded as the twenty most urgently need commodities. For example, they looked at alternatives to the manganese and platinum then supplied by the Soviet Union; they thought Cuba and Brazil could probably cover the former, Colombia the latter (Colombia having been the Allies' main source of platinum in the latter stages of the First World War after Russian supplies were interrupted by the 1917 revolution). Peruvian cotton, they thought, could substitute for Egyptian output and Latin America could supply a range of other materials. The countries of Central and South America had been doing so to Germany and Italy but gradually these links were cut by the British blockade and American political pressure. By 1942 Washington had secured the monopoly on supplies from the region of rubber, tin, manganese, chrome, mica, tungsten, copper, diamonds and lead. Mexico's manganese mining sector, for example, had managed to double its output to meet orders from north of the Rio Grande.

Rubber planting was tried in California but yielded little (apart from farmer anger at taking food-producing acreage way). Then there was the plan to produce synthetic rubber out of alcohol derived from agricultural and forest products. By 1944 rubber production from alcohol butadiene plants had reached 361,734 tons a year and from petroleum 195,719 tons. In addition, Ceylonese output exceeded Britain's needs, so some of that island's natural rubber was diverted to North America.

Meanwhile, there were savage reductions in civilian rubber use. By 1943, this was just one-tenth that of 1941. No tyres for civilian use were manufactured in the first nine months of 1942 but this was reversed when it was realised that the condition of tyres on trucks and buses was a growing concern. Then there was the prospect that cars would no longer be in a condition to be driven due to the state of their tyres.

## IX

Notwithstanding the advent of war, the British whaling fleet continued its operations in Antarctic seas. This was a great disappointment to the Japanese whaling fleet owners; they had expected to have those seas to themselves now that war was raging in Europe. But Britain needed its whaling ships at sea otherwise it would have to spend valuable foreign exchange buying additional oils and fats from the United States.

The Antarctic in 1939 was the source of eighty per cent of the world's whale oil supply, with Britain consuming a third of that—but Britain also bought the bulk of the whale oil produced by the Norwegians and Japanese. The Japanese had been expanding their fleet: in the 1938-39 season, they had six mother or factory ships in those southern waters as compared with ten Norwegian, nine British and six German, all accompanied by catcher boats. In the 1939-40 season, ten British ships headed for the Antarctic along with ten Norwegian, and one each from Panama and the United States. Only the Germans would be missing. That season went well for the British. All their ships got home safely but the date the Germans chose for their invasion of Norway (9 April 1940) was such that the Norwegian fleet was still far from home on its return journey, and the ships were ordered to go to Allied or neutral ports to discharge their cargoes.

In 1939 Britain had applied pressure on Norway to reduce its whale oil exports to Germany. The effect of this was that Germany, which had bought 107,053 tons of oil in 1938, was able to obtain only 13,352 tons in 1939 and 12,360 tons in 1940. Had Berlin delayed its invasion of Norway for a few weeks, it would have got its hands on the 55,000 tons being carried from the Antarctic by the Norwegian fleet. Instead, the oil ended up in British hands; in fact, Britain now found itself with far more whale oil than it could use, so sent some to the United States for storage.

Whaling continued in the 1940-41 season, but Britain suffered a severe blow in January 1941 when the German surface raider *Pinguin* attacked the fleet in Antarctic waters and captured three of the Norwegian floating factories now under British control, eleven whale catchers and 23,626 tons of whale oil which was safely transported to a French port for shipment to Germany (and enabled a surge in margarine production in the Reich).

The whaling industry suffered badly in the war. All floating factories owned by British or Japanese companies were sunk by 1945; they had been converted to transport roles for the war effort and, being slow and highly visible, were easily picked off by enemy submarines. Among the tasks for which the Japanese used their whaling ships was as transports for midget submarines (which could be loaded or discharged using the slipway up which dying whales had been winched). In the four seasons in the latter years of the war, only about 7,000 whales were caught in that entire period—just one-fifth of the mammals killed in the 1938-39 season alone.

## X

In its review toward the end of the war, the League of Nations noted in that, by late 1943,

> the stocks that had been accumulated in the United States out of the rapidly rising domestic production and the substantial imports, had grown so large in the case of most base and light metals that they were considered to be in excess of requirements. The United States Government therefore began to curtail the production of certain minerals by stopping the premia paid to some marginal mines. A number of aluminium reduction plants were also closed down late in 1943 and in 1944; aluminium stocks had risen from 1.8 million short tons at the end of 1942 to 4.9 million at the end of 1943. Action intended to reduce copper production in Northern Rhodesia was taken by the British Government in January 1944, when it announced that it

would cut copper purchases by stages until a total reduction of 20 to 25 per cent was reached.

*Time* magazine reported in early 1944 that the United States had 100,000 tons of tin remaining in its stockpiles, against the rate of use of 50,000 tons a year. This freed Washington of its dependence on Bolivian tin. It was just one example of America's comfort zone in terms of a range of resources.

It was a situation of which Germany and Japan could no longer even dream about.

Not that the Allied side was without hubris: in early 1943, the British Association for the Advancement of Science declared that the anti-Axis nations by then controlled so much of the world's resources that the *Christian Science Monitor* was able to proclaim that 'war can be—probably will be—outlawed by joint control of the distribution of the world's mineral wealth by the United States, the British Commonwealth of Nations and the U.S.S.R'. They spoke a little too soon.

*A U.S. Office of War Information poster issued in 1943.* RPG Nordic.

# 4.

# The Battle for Food

IN 1937 IT WAS estimated that Germany was eighty-three per cent self-sufficient in food supplies. Austria was not in quite such a comparatively comfortable position and Anschluss meant the combined overall Reich's self-sufficiency fell; but once Czechoslovakia was added in 1939, that deficit was more than cancelled, at least on paper. However, self-sufficiency on paper was not quite backed by the reality. The generality of that overall self-sufficiency obscured considerable variations when the food categories were broken down; the figures did not tell the whole truth because there were still serious gaps. For example, there was assumed to be adequate supplies of potatoes and beet sugar, but a serious deficiency in terms of fats. And almost all Germany's maize was imported, with Hungary, Poland, Yugoslavia, Romania and Bulgaria the main suppliers. And on the availability of that maize depended Germany's ability to feed its beef cattle and milking cows. Also with pigs, there was still not enough to go around and pork shortages were made up by imports from Denmark and the Baltic states. Farmers in those two areas were also able to provide butter to fill the gap left by Germany's farmers. Whale oil was sourced largely from Norway. Oilseeds would turn out to be

a problem area: until the outbreak of war, Germany obtained these
from Africa, India and the Far East.

Coffee supplies had to be obtained from Latin America, Africa
and the Netherlands Indies. For smokers, war would pose a real problem: Germany was capable of producing only about a quarter of its
tobacco needs, and was heavily reliant on the United States.

Yet, and in the face of the actual facts, Hermann Göring took to
the radio in February 1940, during the so-called 'phoney' war on the
western front, to boast that Germany's farm economy was invincible.
Nevertheless, he urged farmers to boost production of root crops
and fats (and announced small increases in milk and butter prices to
encourage the latter) and offered subsidies for new acreage brought
into production. Vowing that Germany would never be defeated by
the British blockade, Göring ridiculed Britain for imposing rationing
on its people.

It was another empty boast. As 1940 dragged on, Germany's food
stocks were nearing exhaustion and Berlin was greatly concerned.
Part of the problem was that in the twelve months to May 1940
the Wehrmacht had conscripted, along with men from other key
occupations, around 750,000 farmers and farmhands. There had been
a bumper crop in 1938, but the 1939 one was poor. Short of labour,
farmers eliminated those crops requiring the highest labour intensity
such as potatoes and root crops; the shortages of the latter would have
an impact on the supply of animal feed. Berlin came up with a plan
to replace those three-quarter million army recruits from farms with
workers from Poland, but less than a third of that target number were
actually brought to Germany

But at least part of the solution lay next door. Denmark, after
occupation, had lost its main export market, Britain. In April 1940
the *Boersen Zeitung* newspaper in Berlin boasted that Germany's invasion of Denmark and Norway had, based on 1937 statistics, deprived
Britain of thirty-nine per cent of its imports of butter, sixty-eight per
cent of its imports of bacon, fifty-seven per cent of its imports of eggs.
Before the war, Denmark produced 396 million pounds of butter, of
which 264 million pounds went to Britain, with the remainder split

between other European markets and Danish consumers. Figures produced in *The Economist* in September 1939 showed that, for 1938, 270,000 dozen eggs were imported into Britain from the Baltic and Scandinavian countries—36.6 per cent of all egg imports— of which 95,100 dozen came from Denmark. It took only a few weeks after the Nazi occupation of Denmark before agreement was reached that these food supplies would be sent henceforth to Germany.

Less than a month after the invasion of Denmark and Norway which began on 9 April 1940, the *Chicago Tribune* reported the two countries would fill 'Nazi larders' for months to come. The new German rulers had become masters of an additional 3.5 million head of dairy cows and beef cattle. Norway would add another 1.5 million cows, the Netherlands three million and Belgium two million.

Soon the military forces of Nazi Germany would make available to the war effort the farming land of much of the rest of Europe. Between the second half of 1942 and the first half of 1944, Germany imported nearly 2.28 million tons of potatoes, of which the bulk came from Poland followed by France, the Netherlands and Norway. In 1942-43, 262,000 tons of pork and other meat was shipped from the Soviet Union along with 256,000 tons of butter and vegetable oil and another 256,000 tons of fruit and vegetables. France shipped 1.4 million tons of grain to Germany and 1.39 million tons of hay and straw in that period. The Netherlands was imposed upon for 743,000 tons of fruit and vegetables and 12,500 tons of cheese. In that period, too, Denmark shipped to its large neighbour to the south 42,000 tons of grain, 212,000 tons of meat, 197,000 tons of fish, 90,000 tons of butter and vegetable oils, 7,000 tons of cheese, 4,000 tons of sugar and jam and 12,000 tons of fruit and vegetables.

Of course, three factors separate Denmark from much of the rest of occupied Europe. One, it was allowed to keep its own government until 1943 and had a fair degree of autonomy (even refusing a German request soon after it was invaded to form a customs and currency union) although this came with the price tag of being declared an enemy state by the Allies and, for example, having its merchant ships seized; moreover, it was the only occupied area to be administered by the German Foreign Office rather than by the security and military organs (until mid 1943, that is, when the Wehrmacht imposed martial law). Two, its food shipments were initially much more voluntary in nature than from other occupied countries. Thirdly, no fighting took place on its soil once the Germans moved to occupy the country; the Wehrmacht stayed until the surrender to the Allies in May 1945.

As historian Joachim Lund explained in a 2004 study of Denmark's place in the world order, by 1941 76.4 per cent of that country's exports were going to Germany, three times the level in 1939. Denmark was Germany's second largest trading partner (after Italy). Between ten and fifteen per cent of the food consumed during the war by the Germans came from Denmark. Even then, Denmark was not able to maintain the rate of productivity that had been normal before the war. According to John Christmas Møeller, the former Danish cabinet minister who had fled to London to help the exiled government (and was to become a key broadcaster for the BBC's Danish service for the rest of the war), butter production had dropped from 396 million pounds a year to 264 million. He was reported in *The New York Times* saying his country had also manufactured 176 million pounds a year of margarine but this had ceased as the factories could not get the necessary soy beans. This rebounded on the Germans, because Danish consumption of butter more than doubled in the absence of margarine, leaving less for the Germans to take—although eventually take more they did.

A recent study by Paul Brassley *el al* points to an example of how the Germans' lack of a master plan for running Denmark hampered efforts to supply food to the Reich. In early 1940, ahead of the invasion, the Oberkommando der Wehrmacht was recommending it as essential that Germany supply Danish farmers with feed and fertiliser. As this study shows, the Germans fully realised that Denmark had one of the most advanced agricultural systems in the world. The OKW saw that Denmark would become an important source of food for the Fatherland, and that production had to be maintained.

Of course, this feed and fertiliser recommendation was not put into effect once Denmark was invaded. The lack of imported inputs crippled Danish agriculture. A Danish journalist, Joachim Joesten, had fled to Uppsala in Sweden and filed in August 1940 a detailed analysis for *The Washington Post* in which he explained that Denmark had imported 641,000 tons of fertiliser in 1938, but all it was now getting was about 100,000 tons of potash from Germany and the promise of some phosphate from the Kola Peninsula in the Soviet Union. There were all the other agricultural inputs that Danish farmers and food suppliers needed but were now unavailable: the 308,000 tons of maize that would normally be shipped from the United States, the 718,000 tons of oil cakes, the 35,000 tons of soy beans imported from the Far East, 37,000 tons of feeding flour and 370,000 tons of oil seeds—not, of course, to mention that the gasoline and oil previously used by farmers was no longer available in any quantity.

The Germans knew from the beginning what their invasion of Denmark would mean: the cutting off of farmers in that country from their British suppliers of feeds and fertilisers, with the consequent reduction in animal numbers and crop yields, and therefore less to be available to be shipped to Germany. In June 1939 Denmark has signed a non-aggression pact with Germany, the only Scandinavian country to do so (Finland, Sweden and Norway all declining the offer from Berlin). The Danish foreign minister Peter Munch (who would later be blamed for Denmark's subsequent misfortunes) boasted that, in case of Germany being at war, Denmark's foreign trade would not be affected and could carry on exporting to Britain. The Danes decided the pact was a better bet than a Nordic defence pact as had been suggested by other Scandinavian countries. Indeed, six days after Britain declared war in September 1939 the German diplomat Ulrich von Hassell was in Copenhagen reassuring the Danes they could continue trading with the British provided there were no disruptions to German-Danish trade. Of course the Danes wanted to keep trading with Britain: that yielded them an annual trade surplus with London to the tune of 300 million kroner even after all the purchases of cattle feeds and fertilisers.

The Germans were fully aware that impeding the import of feed and fertilisers would have a knock-on effect as to the Danish food exports so desperately needed by Germany, so there was worked out a deal whereby Danish ships, clearly marked and having provided bills of lading to the Germans, were able to sail to and from Britain without being impeded or sunk by the Kriegsmarine. This system continued until the invasion of Denmark on 9 April 1940 although that did not stop the German navy taking matters into its own hands: five Danish ships sailing under the agreement were sunk in the months before the invasion. But the trade was, of course, stopped when German troops crossed the border. The agreement under which Denmark could trade with Britain had been put in place because there were those in Berlin who fully realised that Germany's food supplies were highly dependent on the ability of Danish farmers to feed their cows and beef cattle with maize, barley and other feeds, and fertilise their land. After all, due to the British blockade in the First World War, the sudden shortage of feed and fertilisers to (the then neutral and un-invaded) Denmark impacted severely the output of its farms. As *The Times* noted just after the April invasion, 'Her [Denmark's] programme of intensive arable dairy farming must also suffer for lack of imported fertilisers which are used freely in Denmark'. The modern Danish agricultural sector was founded on the use of imported grain and fodder. Again, Germany ignored its own economic interests in pursuit of its military desires, a choice that recurred throughout the struggle and helped doom the Nazi war plan.

The failure by the German authorities to provide feed and fertilisers had the predictable effect: by 1942, butter production had fallen by forty-two per cent since 1939, and declined again in 1943. By 1942, pig-meat production in Denmark was just one-third of the pre-war level. The only ameliorating factor was that Denmark had good grain harvests for much of the war.

Yet some sense did prevail. The Danish fishing fleet was allowed by the Germans to go to sea. Huge catches were made even though diesel was in short supply, and the fishing areas limited to areas that were not mined or designated as a military zone. Germany benefited enormously. In 1938, it had imported 20,630 tons of fish from Denmark; by 1943, the figure for that year was 99,281 tons.

Even though the Danes could not supply as much food as Berlin would have wanted, enough was left for Danish mouths which contributed to the country maintaining price stability through 1943 and 1944, while inflation was roaring in the rest of Nazi-occupied Europe (largely because Denmark had been integrated into the total German food market, unlike the other occupied territories). The other main factor was that Denmark was able to manage its own affairs, and the authorities in Copenhagen increased taxation through a war boom tax, an excess profits tax, increased indirect taxes and compulsory savings. Money supply increased nevertheless, but inflation did not get out of hand.

After Denmark, the German planners began considering the Ukraine question. The view was that the capture of this 'bread-basket' was vital to Germany having enough food.

## II

The invaded territories put under most pressure to supply grain to Germany were France and Ukraine. But their degrees of suffering as a result were not comparable.

The Nazis saw Ukraine as the solution to their food problems. In 1942 the agronomist and longtime Nazi Dr Herbert Backe took over as Reich Minister for Food and was the architect of what would later be called the 'hunger plan', essentially the decision to condemn thirty million 'useless mouths'—the urban populations of the bread-basket regions of the Soviet Union (Ukraine particularly)—to death by starvation. The Backe plan called for the food normally grown in these regions, which sustained the great cities of the northern regions over the long winter, to be diverted initially to feed the Wehrmacht and German civilians. As Lizzie Collingham shows in her study of food issues during the war, Backe at least among the Nazis realised the problems of supplying enough food. He had risen through the Nazi ranks, pushing the hunger plan while state secretary at the Reich Ministry of Economics and then given full control when he became food minister. His argument was that German agriculture could not provide sufficient food to feed an army which by then was home to about one-seventh of the German population (and which, as shown elsewhere in this book, had absorbed a large proportion of the farmers and farm labourers). The Soviet Union would have to provide the entire amount of food consumed by Germany's military forces. The Nazis faced a situation in 1941 where Germany could not feed itself: the only solution which presented itself was to seize the lands of those who could be forced to feed the Reich. As with oil, if the Wehrmacht could not secure the food resources, then Germany could not win the war. Talk about rolling the dice.

And luck was not a lady in 1941. As Collingham details, the 1941 Ukrainian harvest was far lower than expected. Not only did no food get shipped back to the Reich, the troops in Russia (and especially those trying to fight their way into Moscow) did not have enough to eat. Luftwaffe planes strafed peasants working in the fields; peasants had destroyed much equipment as a protest against forced collectivisation. Moreover, many farm workers were able to hide their harvest before the Germans arrived in their areas and, where the Germans were already in occupation and supervising the gathering of crops, troops were often called away to meet Russian attacks, so leaving the locals to also hide much of their food.

In October 1941 a Swedish journalist, Helge Knudsen of Stockholm's *Dagens Nyheter*, journeyed through the Ukraine. The report detailed how the Soviets had removed much farm equipment eastwards to prevent it falling into German hands; that which could not be moved was disabled. 'Along the roads of the Western Ukraine the Germans found incredible numbers of tractors, often as many as two per kilometre. All had been made useless by the removal of vital parts', read the translated account as it appeared in *The Scotsman*. Knudsen said he witnessed Germans trying to harvest corn using scythes, sickles and with old reapers drawn by horses. The report stated that farm workers had, under the Soviets, been paid as wages every eighth sheaf they harvested, but the Germans did not have enough corn to follow this and instead decide to pay in cash. As Knudsen pointed out, as there were no goods to buy in the occupied territories, the peasants had no interest in working for the Germans.

The hunger plan had one significant economic flaw: instead of turning up for work in the factories of the captured Soviet territories, the workers spent their time foraging for food. Not only that, but the Nazi plan swelled the ranks of the partisans, thus diverting troops away from the fighting front.

Meanwhile, in occupied France there was, like elsewhere now under Nazi control, food rationing and the Germans were determined to enforce the rules. Food was soon in short supply in France and, when it was available, the cost was often prohibitive due to soaring inflation; the shortages problem initially was exacerbated by the understandable inclination toward hoarding after the shock of the defeat of France. Most French people were suddenly faced with a situation where they could not longer obtained those items they had long taken for granted: cheese, for one; eggs were often scarce (and soap usually unavailable).

It was estimated in 1942 that the French black market was worth somewhere between 60 billion and 70 billion francs. By the summer of 1940 French agriculture had become severely depleted of manpower, fertiliser, horses, transport and fuel. The policy of imposing forced labour in Germany for men exacerbated the problem caused by so many farm workers having signed up for the army and now being prisoners of war. Those remaining on France's farms not always engaged in the black market for financial profit, but often out of need, trading food for such hard to get items like shovels. City people with relatives or friends in the country were getting food parcels; in 1942 it was estimated the average person in occupied France was getting 1,725 calories a day under the ration system, 200 calories from food bought on the black market and another 200 from food parcels.

Lynne Taylor, an associate professor of history at the University of Waterloo in Ontario, has written widely about occupied France. She cites several examples of the scale on which black markets operated: in June 1944 one butcher was caught in possession of 1,479 kilograms of pork, seventy kilograms of flour, thirty kilograms of sugar, three pots of edible oil, along with 170,000 francs in cash. Another, who had two vans, was arrested after authorities found him with seventy-nine kilograms of butter, forty-nine eggs, two pots of edible oil and 252,000 francs in cash.

> As required by law, the butchers opened their shops Wednesday mornings and sold their stock at the official prices for one hour. Then they habitually would declare that they were out of stock. A little later in the day, and on subsequent days, they would sell rabbit at seventy francs per kilogram and fowl for 100 francs per kilogram, which were extortionate black market prices.

In 1941, police at Lille investigating a garage found 480 kilograms of chocolate, two thousand kilograms of canned peas, one thousand kilograms of canned meat and 237 kilograms of pepper. The following year, Taylor recounts, a check of all passengers arriving at Lille railway station ended up yielding quite a pile of clandestine goods: 9,550 kilograms of potatoes, 170 kilograms of beans, 120 kilograms of peas, 430 kilograms of wheat, 230 eggs and sixteen chickens.

But, of course, it was not only the hungry French who were using the black market in that country. So were the Germans, both the various agencies and also individual servicemen. Some black market operators made fortunes working for the Nazis. One of the most successful was Joseph Joinovici, a Romanian Jew who arrived in France in 1925 and got into scrap metal. He also had the foresight to spend some of the proceeds on forged documents showing he was of Aryan descent. It has been estimated that he made fifty billion francs in the four years of occupation through delivery of scrap metal to be shipped to Germany. Purchases on the black market by German organs in 1943 alone amounted to an estimated forty-four billion francs. Soldiers supplemented their monotonous rations by buying more appealing food from local black marketeers.

While the Germans were constantly frustrated by both the police turning a blind eye to petty black market activity and the courts being lenient with the more serious offenders in the occupied zone, the Vichy authorities were keen to show they had more resolve. The Vichy government estimated that fifteen per cent of food being consumed had been traded in the black market. In August 1942 fines and sentence lengths for those convicted of black market activities were doubled. The *Christian Science Monitor* quoted *Le Temps* in October 1942 complaining that food profiteers had expanded their influence to equal the marketing power of the bootleggers during the prohibition period in the United States. But unlike the American situation, which involved only alcohol, the black market in Vichy France enveloped almost every aspect of life, with the underground businesses trading in butter, eggs, cheese, ham, fowls, coffee, tea, tobacco, cloth and even shoe leather.

But there is another side to the story. As Paul Sanders in his study of French black market shows, the dislocations of the war by themselves caused scarcity and, as night follows day, higher prices. By July 1940 it was estimated that retail prices had risen by 150 per cent. And while the criminality of profiteering was only too rife, the black market probably prevented a supply disaster in France. A correspondent for the *Milwaukee Journal*, Melvin K. Whiteleather, filed a report from Algiers in April 1944 quoting people arriving from France telling him the black market was a 'blessing', that they would not have eaten unless the farmers had saved the day by cheating the Germans. 'If a producer could hide away one or two pigs or a few dozen eggs or several bushels of grain or a quantity of vegetables, he did so,' the report noted. And the reporter added: the producer was making money but it seemed the French were prepared to forgive.

The problem was Germany could not keep all its balls in the air. It starved the occupied territories of petrol in order to meet its own needs, yet this had knock-on effects elsewhere, including with food supply. After the French surrender, occupied France was reduced to existing on less than ten per cent of the petrol it had consumed hitherto; one consequence was that a large proportion of France's milk production was poured down the drain because the trucks had no petrol with which to make collections.

Apart from the disruptions of the fighting and occupation, the food situation was further complicated in years of poor harvests. As the League of Nations noted in its 1945 economic summary, grain harvests in Europe during the war were consistently below the 1933-37 average. Grain crops were very poor in 1940; a little better in 1941, and again very poor in 1942. Efforts were made to increase plantings of potatoes, sugar beet and vegetables to help make up for the grain deficiency, not entirely with success. Livestock numbers suffered due to fodder shortages. 'By 1942, the production of both vegetable and animal foods had dropped very considerably below the pre-war level, reserve stocks had been largely depleted, and imports of foodstuffs from overseas were reduced to a mere trickle,' the league noted.

Sweden, being neutral, was still able to import under the 'safe conduct' arrangements negotiated with both Germany and the British. Hence Swedish livestock herds increased in 1943, and oil-bearing seeds were also being imported. For the other neutrals, the story was not so good. The Swiss were having to get by on less food than that to which they had been accustomed before the war, and severe droughts in 1943 ruined much of the Portuguese and Spanish harvests (in the latter's case, exacerbating a situation where food production had still not recovered from the dislocations of the civil war).

Overall, European agriculture was constantly hampered by shortages of labour, fertilisers, horses to pull equipment and machinery generally. Yet there was a silver lining for some farmers. The unavailability of all the necessary agricultural inputs, along with the lack of manufactured goods to buy, meant farmers were left with considerable cash balances. Many used this surplus to pay down their debts. As the League of Nations charted,

> In France, Hungary and elsewhere on the European continent there was a marked improvement in the financial condition of farmers as compared with other classes of the population. The weight of agricultural indebtedness was substantially alleviated in countries as wide apart as Canada, Denmark and India. In India especially, the inflationary price rise in 1942-43 was not without

some beneficial effect, considering the heavy debt burden which had weighed on the Indian peasantry previously.

The Americans felt aggrieved by their sacrifices. In 1943 almost every city was reporting shortages of meat but you could still buy a leg of lamb. You could still go to a restaurant and be served a nourishing meal, even if many days it was a meatless one. Grocers rationed canned goods, you needed coupons to buy sugar (due largely to the Japanese capture of the Philippines and the lack of shipping space to haul sugar from Hawaii), butter and vegetable supplies could be patchy. Coffee, too, was sometimes hard to get. But the American 'hardships' were bearable, not so in Japan and Italy and later Germany. After all, some of the shortages afflicting the American civilian population were due to the need to ensure plenty of food was available for the fighting forces and for supplying Britain under Lend-Lease.

### III

The weakest link in the domestic economic chain of Japan was agriculture. Much of the home islands was rugged in terrain and the soils on the flatter land were often not good. Farming families had to play their part in the great manpower effort of the war with millions of young men needed not only for the armed forces but for vital wartime manufacturing. And Japanese farming was, even by international standards, very labour-intensive. On top of that, fertilisers were being diverted for the making of explosives while farm horses were sent off to pull army equipment in the Chinese theatre.

In 1939 and 1940 measures were introduced to control rice distribution in Japan; these rules were then extended to wheat and barley.

It was hard enough to keep the Japanese fed, so it became imperative that each territory captured by imperial forces play its part in feeding the vast armies and other forces operating throughout Asia. When the Philippines were taken, the notion in Tokyo was that the islands could support Japanese troops as well as feed its own people. But it was a misconception; much of the Filipino farming effort was ordered by the Japanese to be diverted into cash crops, many for export, such as abaca (a banana-type plant whose leaf stalks were used to make hemp), sugar and copra. Often, indeed, the Philippines needed to import foods; the focus on export crops had come at the price of not having food self-sufficiency. Moreover, much of the country's potentially arable land had not been cultivated and irrigation was often rudimentary, if it existed at all.

The wartime administration in Manila promoted, as much as it could, efforts to grow more crops and bring idle land into production; municipal authorities were to distribute seeds and cuttings for planting. The Japanese did help by bringing in better rice strains from Taiwan and having these planted on a wide scale, and they trained farmers in the double-crop system and added another 100 days to the average 160 days a year to which the Filipino farmers were accustomed to working.

But it was not enough. Rice had to be shipped from Saigon in 1943 to fill the supply gap, and starvation was by now occurring in the Philippines due to the small per capita rations. In addition, the army was still taking large quantities of food which would normally have made their way to the civilian population. Moreover, part of the Philippines sugar crop was diverted into making butane for aviation fuel. But the Japanese had made a problem for themselves as well as the Filipinos: by forcing the conversion of sugar lands into cotton growing, by the latter years of the war there was not enough sugar to supply the fuel-making processing.

In China, the plan was to disrupt the silk trade—silk being seen as an unnecessary luxury in wartime—by uprooting mulberry gardens and replacing them with food crops to feed the Japanese armies and the home islands. This followed a gradual process in Japan itself in the 1930s where cocoon and silk production had been run down, a process that speeded up from 1937 when the war effort in China exacerbated labour shortages. A final clamp-down was ordered in 1940 under the luxury prohibition scheme; and exports to the United States fell away as nylon made inroads into the hosiery industry at silk's expense.

In Singapore, no open ground was left untilled in as the people strove to maintain adequate food supplies under the Japanese occupation (the Japanese having renamed the British colony Syonan-To). The Grow More Food Campaign was launched in the expectation that Singapore could be blockaded by the Allies, and therefore had to become self-sufficient in food. Food shortages were severe during the occupation, and tapioca (favoured because it grew well in various soil conditions and matured in just three months) along with sweet potatoes and yam became the staples of survival for the local people. The Japanese authorities issued ration cards to control supplies of rice and other essential items; they also reduced rations when they believed the locals were not doing enough to produce crops. By 1944 the Japanese became more severe in coercing people to plant food by further cutting rations of rice, sugar, salt and coconut oil. In December of that year, men in non-vital occupations (such as tailors and shop assistants) were forced to become farmers.

However, it seems that it was largely the labouring class that grew the food; many city dwellers resorted instead to the black market. The lack of fertilisers was overcome by using human excrement to enrich the soil—a practice that lead to the spread of typhoid, dysentery, cholera and intestinal parasites. The year 1937 had seen two hundred Singaporeans die of dysentery; that figure reached 2,977 for 1944.

People planted crops wherever there was an empty space. A children's playground was turned into a vegetable patch, and inmates at Outram Prison planted crops in an unoccupied area outside the fence. Even the Padang Cricket Ground was planted with tapioca trees. The Japanese published a recipe book showing how tapioca and other native food could be prepared. Across the Straits of Johor in occupied Malaya, forest reserves and rubber plantations were cleared for cropping.

Surprisingly, perhaps in view of its large land area, Malaya in the years between the wars managed to grow only about a third of its rice needs. The staple was made a prohibited item under the Japanese occupation: all stocks of rice had to be declared and collection of crops was done under official supervision. In March 1942 the price was fixed at seven Malay dollars a picul (a weight measure used widely in South-east Asia and China and equivalent to about 60.5 kilograms). By 1944, rice's price on the black market had soared to $480.

## IV

And then there was the case of Vietnam. Somewhere between one million and two million people starved to death in the final year of Japan's occupation of what is now Vietnam. That is the view of Cambridge University's Bui Minh Dung. He considers all the potential contributing factors in the chronic shortage of food—the Allied bombing, bad weather, French policy, inflation, permanent subsistence crisis, social inequality, local hoarders and speculators— and, while acknowledging these and the fact that each one served to exacerbate the effect of the others, he has no hesitation in concluding that premeditated Japanese policies were ultimately to blame. There was bad flooding, the French authorities themselves hoarded rice in their own depots, inflation was out of control due to scarcity, but it was all made worse by Tokyo's intervention. After Japanese troops arrived in Indochina, there was forced crop conversions. The Japanese wanted more rubber, more cotton, more edible oils and more fibre crops such as flax and jute. In the north of the country, the peasants were told to shift from growing subsistence plants including maize to producing cotton, jute, hemp, peanuts, caster oil and sesame.

## V

But, if the people of the conquered territories did it tough in relation to getting enough food to survive, then the Japanese civilians at home were pretty much in the same boat. Or, in fact, they were not able to be in the boat at all: most of the protein consumed by Japanese in that era came from fish. But the fishing industry was one of the first to suffer from the war effort. The labour force on the boats and shore processing fell from 1.1 million in 1936 to 721,000 by 1945, due to a combination of men being conscripted into the army, the government requisitioning all the large fishing boats for military use, and then the acute shortage of fuel (by 1941 the fishing fleets were getting only about one-third of their prewar supply of fuel oil) and fishing nets meant that many boats still with crews could not actually go to sea, or not all the time, anyway. The fish catch fell by a half, a serious dietary blow for people on the home islands. According to a U.S. Department of Interior, in 1939 Japan had 71,641 powered fishing vessels (there were another 283,000 unpowered ones), but of those only 125 had steam-driven propulsion; the remaining 71,516 ran on diesel or petrol, all of which was imported.

On land, scarce grain needed to be given to humans, leading to a huge drop in the production of chicken and pig meats.

Nature deal the Japanese a few blows, too. In 1939 a severe drought hit Korea and the home islands. Before that, Japan relied heavily on Korean rice but little was available that season, and stocks in Japan fell by half.

By 1943 there was little shipping space available to carry rice from Southeast Asia back to Japan. In the previous twelve months, Indochina and Thailand between them supplied seventy-four per cent of supplies in Japan; the following year, shipping space was so short that those territories accounted for only ten per cent of Japan's rice needs, with the Japanese reverting to heavy reliance on Korean rice; rice (and sugar) supplies from Formosa had by this time been disrupted by an Allied blockade of that island.

Of course, little was done ahead of the war and, when something was done, it was always too late in Japan's case. No better example is the move in May 1945 by Japan's Minister of Agriculture, Tadaatsu Ishiguro, who called for a food self-sufficiency program that would sustain the country for the next twenty to thirty years of war. Baron Ishiguro had been one of the leading agricultural officials from 1919 into the 1930s, and he donated part of his salary to assist tenant farmers. This was his second stint as Agriculture Minister, first having served in the Konoe cabinet (1940-41). This 1945 broadcast was interpreted as indicating that Japan was preparing for dislocations of the food supplies coming from Manchuria and Korea. This came as radio broadcasts told Japanese that their country was being divided into blocs so they could organise an area by area defence of the homeland in time of invasion.

Too little, too late—the recurring theme of Japan's economic management.

# 5.

# The Advantages of Empire

THE GERMAN COLONIAL LEGACY was quickly forgotten in most of that country's pre-1914 imperial possessions—German Samoa, Tsingtao in China, the islands of Micronesia in the Pacific, Togoland, German Cameroons, Tanganyika and New Guinea. These were now all being run by other powers: New Zealand (Samoa), Japan (the Micronesian territories), Australia (New Guinea) and Britain (Tanganyika) had since imposed their systems usually under League of Nations mandates. And, anyway, the period of German rule had been short and, with one exception, was not accompanied by any notable settlement by German colonists.

By contrast the Germany legacy seems to have survived a remarkably long time in the former German South-West Africa (now Namibia). One of the reasons for this may have been that, contrasted with their other territories, the Germans had left quite a mark on South-West Africa. The huge diamond discoveries had led to a dramatic transformation, with considerable economic development and private investment. By the time the Germans were kicked out, the territory had 2,100 km of railways, including a link to the South Africa rail network.

In May 1942, a *Time* magazine correspondent visiting a territory that by this time was administered by South Africa was astonished to find so many leftovers of German rule. He remarked that one of the main streets in Windhoek, the capital, was still called Göringstrasse, the implication by *Time* being that it honoured the Nazi leader; in fact, it was named after Hermann's father, Heinrich Ernst Göring, who had been the high commissioner there between 1885 and 1900 (the street was not renamed until about 1990). The reporter took beer and apfelstrudel at the Café Vaterland and was offered the daily newspaper, *Deutsche Beobachter*. Travelling to Swakopmund, he found the main hotels to be called Hansa, Fürst Bismarck, Thüringa Hof and the Europäischer Hof. As with Windhoek, the main street was still the Kaiser Wilhelmstrasse.

## II

In September 1939, troop numbers for the entire British colonial military force throughout Africa stood at 40,000, that number including the King's African Rifles, the West Africa Frontier Force, the Sudan Defence Force, the two Rhodesias and the high commission territories Basutoland (now Lesotho), Bechuanaland (now Botswana) and Swaziland. By May 1945, with the end of the war in Europe, more than 450,000 men from these British African colonies were in uniform. If you set aside the Indian Army, about eighty per cent of Britain's colonial military manpower came from the African continent in World War II. They served not only against the Italians in East Africa, and the Germans in North Africa, but in Madagascar, Palestine and Burma. As David Killingray writes:

> Apart from the large standing Indian Army, Britain's colonial forces had been allowed—largely due to the depredations of the Great Depression—to fall into pretty bad shape during the 1930s. However, improvements in communications also contributed: the importance of a local, standing force was lessened by the ability to summon aid by cable and despatch troops by fast motor ships.

The Italian invasion of Abyssinia certainly concentrated minds in London in respect to the African colonial forces at its disposal. Suddenly the security of the Suez Canal and oil supplies from the Middle East were of concern. In 1936, there had been a tussle between the War Office and the Colonial Office, the former wanting to gain control of local forces until then administered by the latter. Eventually the imperatives of the war, including the Japanese entry and the threat to India and possibly Ceylon, gave the War Office control over African units. In fact, once the Vichy threat in West Africa had ended, there was agreement that African forces, which had fought well against the Italians, should be used outside the continent. About 120,000 were stationed in Ceylon, thus freeing British garrisons there to be deployed to the front.

Before the war, however, just how useful were the colonies? After all, there were plenty of them. According to a count in the magazine *Picture Post*, there were some 120 such territories before the war, containing a total 250 million people; this clearly did not include India, which in 1941 contained 318 millions, and which was administered from the India Office in London (rather than by the Colonial Office) and had its own cabinet minister. The British non-India territories contained sixty-seven millions of the 250 millions figure, with the Netherlands in second place with sixty-four million colonial subjects. In area terms (and, again, obviously excepting India) France ruled the greatest amount of ground. Japan had thirty million colonial subjects (Korea and Taiwan, along with the trustee islands off Micronesia), the United States fourteen millions (into which were lumped Alaska and Hawaii as well as Puerto Rico), Belgium thirteen millions and Portugal nine millions.

Emanuel Moresco, a former vice-president of the Council of Netherlands India (also known as the Dutch East Indies, now Indonesia), had become somewhat of an expert on colonial questions, leading to the publication by the League of Nations in 1939 of his book *Colonial Questions and Peace*. He was sceptical about the economic benefits derived from having colonies. Moresco believed the British empire gained far more from its association and trade with Ireland (then called Eire) and the Dominions (Canada, Australia, South Africa, Newfoundland and New Zealand) than it did from all its colonies, territories and mandates. In fact, Britain was little better off than other European powers (Moresco excluded Russia): the British and the peoples of the Dominions still had to source quite a few raw materials from beyond their collective borders, especially petroleum, potash, antimony, mercury, silk, flax and hemp. Even with her colonies, he argued, France was lacking in basic materials in sufficient quantities such as coal, rubber, petroleum, and cotton. Italy's colonial possessions in Africa were even less help, with the metropolitan power dependent on imports from outside itself and its imperial possessions for all its rubber, lead and nickel and practically all the cotton, copper and mineral oils, along with coal and wood pulp. Italy did manage to supply twenty per cent of its linen requirements and forty per cent of its iron, steel, wood and graphite needs.

Hitler frequently referred in the pre-war years to the loss of German colonies after the 1918 defeat; as late as February 1939 he was still including in his speeches claims that colonies were a necessity for the German economy. As I have shown elsewhere, these German colonies were never of great importance to the metropolitan power, and they never had the allure that British and French possessions offered to their metropolitan powers. In fact, the Germans outside government and certain commercial circles pre-1914 took almost no interest in their empire. As I wrote in my book about naval action in the Indian and Pacific oceans in the Great War, *German Raiders of the South Seas*:

> As economic conditions improved after 1900, emigration from Germany slowed to a trickle. Moreover, getting immigrants interested in the new German colonies was not easy: there were no temperate ones with large swathes of potential farmland, or anything vaguely approaching the appeal of the Cape Colony, New Zealand, Canada or the Australian colonies. The Cameroons and Togoland were seen as tropical hellholes, and South-West Africa was unsuited to farming because so much of it was arid.

By 1913, this entire empire was home to just 23,500 Germans, and many of them were serving in the administrations, army or police forces rather than as people making a new home. This lack of critical mass of Europeans in the German colonies also meant these territories never became a meaningful market for manufactured goods from the home country.

And, unlike with British colonies even after they began winning their independence in the 1950s, the German colonies post-1918 had little residual value for either side. True, after the First World War the ties were cut suddenly because Germany's colonies immediately became someone else's (although ostensibly administered as League of Nations mandates) but Germany went into the Second World War with few loyal, or helpful, friends around the world. In 1937, the newspaper *Berliner Tageblatt* studied the trade that year between Germany and its former colonies. These territories were no more productive by the mid-1930s than they had been under German rule: apart from sisal from Tanganyika, phosphate from Nauru and vanadium from South-West Africa, the former colonies played no significant role in German (or any other) trade. Moreover, once the Nazis gained power and their military inclinations became obvious, neither Britain nor France would have tolerated Berlin being handed back territories that could serve as naval and air bases outside Europe.

Nevertheless, for Britain's war effort, access to its own colonial resources, along with those of France and Belgium, would play a critical role. For example, when you take the case of tin, Britain and its empire dominated world production with mines in Cornwall complemented by the huge output in Malaya along with production from Nigeria, Australia, Burma and South Africa. Imperial production in 1937 stood at 110,000 tons, against British consumption of 18,000 tons. Even with the loss of mines in Malaya and Burma once Japan entered the war and temporarily occupied those territories, the British war effort still had alternative tin resources on which to call. When it came to nickel, Canada dominated the world production terms, and New Caledonia—the other important source of the metal—remained free for the duration. (New Caledonia became the main United States operations centre for the South Pacific with 60,000 American and 20,000 New Zealand troops stationed there at one point.)

British Guiana (now Guyana) became a critical source of bauxite, and both Britain and Canada had plentiful smelter capacity to produce aluminium. Canada—again—and Australia had long been producers of copper, and these were joined before the war by Northern Rhodesia (which also yielded cobalt as a by-product); this colony also had begun producing vanadium. Lead could be acquired from Australia, Canada and Newfoundland; India produced large quantities of manganese for steel alloys, in addition to which mines were developed in South Africa, Gold Coast and the unfederated Malay states.

Moreover, the network of colonies was important for other than their domestic resources: they provided secure staging posts along the world's trade routes in an era when vessels needed bunkering on a frequent basis, and when money was to be made by ships operating 'milk runs' along coasts, flitting from port to port. In Britain's case the network of ports was of a geographical complexity under one country's control never before seen: Gibraltar, Malta, Port Said and Aden on the route to India; ports along both of the African coasts; ports in Canada and the Caribbean across the Atlantic; Singapore and Hong Kong (and, until 1930, Weihaiwei) in east Asia; and, of course, Australia, New Zealand and Fiji in the Pacific region.

But France's territories also provided a strategic positioning along the world trade routes. Apart from the important ports in North Africa—Casablanca, Oran and, in Tunisia, Bizerte—there was Dakar in French West Africa, Diego Suarez in Madagascar, Djibouti at the entrance to the Red Sea, Pondicherry in French India, Saigon and Haiphong in Indochina, Noumea and Papeete in the Pacific and Fort-de-France in Martinique.

Britain's empire was invaluable for one vital British need: food (for the British and their forces, if not so much for the colonial subjects themselves). And to provide those priority groups with food, the subjects of the empire would be required to make sacrifices, whether they liked it or not. Britain made sterling efforts to feed its own people during World War II. Its record in the rest of the empire was less impressive.

Many colonies were ill-equipped to feed their people even before the war began due to imposed colonial farm production policies. In Tanganyika food production had been reduced in order to expand production of cotton for export. Moreover, much of the peanut and sesame crops (both good sources of protein) were tagged for export. Millet and sorghum, also high in protein, were replaced for domestic consumption by low protein (but high in carbohydrates) maize and cassava. The result: chronic malnutrition and recurring famines. British Somaliland lost sixty per cent of its sheep, goats and cattle during the drought of 1943, and ten per cent of its camel stocks.

The dislocations of the war, combined with the natural disaster of droughts, saw even non-combatant African territories suffer serious food shortages due to drought. As John Laband recounts in his study of civilians in wartime Africa, it was estimated that 25,000 people died due to the drought that gripped Portugal's Cape Verde Islands between 1940 and 1943. Once France fell, the people of North Africa had nowhere to turn. In Algeria, there were great suffering during the drought with an estimated three-quarters of the territory's sheep succumbing; in Morocco, it is believed about half of all sheep died of thirst.

Apart from national disasters, there were economic ones. As Laband recounts, price inflation was one result: in Lagos, for example, the cost of cement, bicycles, corrugated iron and sewing machines shot up. Coastal Nigerians had long depended upon canned food as a supplement to what they themselves could produce. Supplies of tinned food dried up, as they did of kerosene used for lighting. The commandeering of lorries by the army in Uganda meant growers could not transport bananas to the towns, causing shortages and pumping up prices.

Even when food, drink, tobacco and cotton clothing was available, it often attracted punitive import duties as colonial administrators tried to ensure their territories were self supporting. In 1941 the

Colonial Office report, *Economic Survey of the Colonial Empire*, showed that import duties imposed in British Guiana were adding about twenty-five per cent to the cost of imported goods. The West Indies territories were reporting an annual income of £8.32 million pounds, against outgoings of £7.43 millions.

A report after the war published by the Royal Institute of International Affairs showed up the repercussions of the conflict to the French colonies. By 1942, imports into these colonies of coal and petroleum products had virtually ceased, as did those of road and rail vehicles. The territories, because farmers had abandoned much of their food production for the more valuable cash crops (such as groundnuts in Senegal), were reliant on rice imports. Suddenly, these fell from as much as 60,000 tons a year to around 1,000 tons.

Across in Asia and nine months before the Japanese struck, it seemed that Britain's Asian colonies were actually prospering from the war. Tin and rubber profits poured into operations in Malaya, tea and rubber buoyed Ceylon while British Guiana saw good oil profits and Trinidad's bauxite was in heavy demand. But the reverse of the coin was that the colonies which relied on agricultural products were faced with sudden contractions in the market for their produce; coffee growers in east Africa suffered, as did copra producers in the Pacific islands, banana growers in Jamaica and cocoa and palm oil plantation operators in West Africa. London made a good deal of the fact that it bought the whole cocoa crop of 1940 in the Gold Coast and Nigeria at prices higher than had obtained in 1939, but this was small comfort to the growers as their receipts were still lower than they had been in 1931 before the worst of the Great Depression was felt. In fact, Britain did not buy the entire output of cocoa: the mid-crop had no taker and was destroyed, and then in 1941 London sliced twenty per cent off what they paid for that year's crop at the same time as raising the price of the cocoa to wholesalers within Britain by ten pounds per ton, the extra margin accruing to the government. When it came to palm oil, Britain did not need the entire output from Sierra Leone and Nigeria. Those farmers in Nigeria who grew oil palms as part of a diversified operation (and therefore did not rely solely on this crop) were told their output would not be needed, so they should switch to other crops (cassava was one suggested, as Britain needed 10,000 tons of that). *The Economist* estimated that palm oil growers who could not sell their palm oil lost, in total, about half a million pounds. As for cocoa crops, African officials had recommended to London that some extra payment be made to compensate for the fact that growers no longer had an open market. This fell on deaf ears at the Colonial Office, the price being set at £16 10s a ton when growers could not make a viable living at much less than £25. The London response was that, with Germany's market closed off and American demand declining, the growers had no option other than to take what London was offering. It was not until 1944 that prices were raised, by which time many cocoa growers had walked off their land.

Pacific island copra was subsidised, but Jamaica's bananas had been banned from being imported into Britain, primarily because there was no shipping space available. Bananas required a good of space as they had to be hung in special compartments aboard ship. This was a huge blow to the Jamaican economy at a time when tourism had dried up: export earnings from the crop had been £2.4 millions in 1939 but only £172,149 in 1943.

Coffee was in trouble even before the war: not only was global production more than twice the amount of coffee being consumed, the Americans favoured Latin American producers, Australia got its coffee from the Netherlands Indies (although not for much longer) and Britain had stockpiles enough to last for two years. Citrus growers in Cyprus and Palestine had a ready market for their oranges and other fruit in Britain, but not the ships to transport them. George Lloyd, who as Lord Lloyd was Secretary of State for the Colonies (having been ennobled in 1925 after serving as Governor of Bombay; he would die in office on 4 February 1941) said the affected people of the colonies would have to do without what he termed semi-luxuries, among which were counted tinned meat and fish, flour, piece-goods and household items. And, as *The Economist* pointed out in 1941, the consequent decline in imports of manufactured items exacerbated the finances of various colonies as import duty was one of the revenue mainstays.

In fact, London was prepared to inflict serious economic harm on its colonial subjects should the need arise. In Southern Rhodesia in 1937 the legislative council had passed a bill to establish the Cold Storage Commission (CSC) to ensure the viability of the colony's beef industry. It would have a deleterious impact on African cattle farmers. But, before that played out, Southern Rhodesia's Prime Minister, Godfrey Huggins, decided that native labour had to be recruited compulsorily. As described by South African academic Nhamo Sanasuwo, an average of 11,408 African male conscripts were 'press-ganged' every year to work on farms or in mines. Additionally, many squatters, including those on vacant Crown land, were pushed off to make room for new settler farms. Then came the problem of adequate beef supplies. Neither the CSC nor Liebigs Meat Extract Company were able to get enough cattle to fulfil orders for the supply of tinned meats to the empire's armed forces. Regulations were passed to authorise a culling program in the African reserves, forcing the native owners to sell their beasts to the CSC at very low prices. It worked: deliveries of cattle to the commission increased from 27,000 head in 1942 to 100,000 in 1945, as Sanasuwo found.

Swaziland was administered from the office of the British High Commission in South Africa. And the protectorate first got drawn into the war because South Africa needed military labour. Chiefs and headmen did the recruiting with the initial draft being for men to act as guards. But, eventually, South African needs for labour were overtaken by Britain's requirements for recruiting in its own colonial territories. In 1941 Britain established the African Pioneer Corps, drawing labour from the high commission territories in southern Africa, Mauritius, the Seychelles, East and West Africa. Many were transported to the Middle East, but initially they were not allowed to go into towns for fear they would 'misbehave'. In addition, there is evidence that many men volunteered in the mistaken belief they would be serving in a fighting capacity; had they known they were for labour squads, they would have been less enthusiastic.

By 1942, however, coercion rather than volunteering had came to dominate the recruit of Swazi men. This was part of an effort ordered by London to raise another 150,000 men from Africa for labouring work. In Swaziland, there were cases where men were threatened or even abducted by recruiting agents.

But, if the British colonies left something to be desired, the Spanish and Portuguese empires were ruled with frequent brutality by metropolitan regimes under Franco and Salazar. In Mozambique, for example, workers in the timber industry were not even supplied with hand-saws; instead, they had to work painfully shaping wood with adzes. Natives were forced into labouring on rapidly established cotton plantations in Mozambique, the sector's workforce surging from 90,000 to 800,000. Forced labour was common in these colonies: in São Tomé and Príncipe, children under eight were drafted, pregnant women were forced to work, corporal punishment widely employed and some women pushed into prostitution. Moreover, the parlous state of both the Portuguese and Spanish economies, and their dependence upon foreign shipping, meant they could not fill the gap in their colonies left by the withdrawal of goods that had previously been supplied by the belligerent nations.

## III

Cash crop farmers in the colonies were often suddenly thrust back into subsistence agriculture when they could not send away their produce. This was felt particularly strongly in the French territories; many of the African ports, particularly Algiers, had always depended on foreign trade to thrive. All but Chad of the French African territories declared for Vichy. Therefore, the others were subjected to the British blockade, depriving the farmers of their usual markets in France for groundnuts, cocoa and rice. Almond merchants in Algeria had long lived well; now they found their businesses going broke. When the grape harvest was brought in by Algerian growers at the end of 1940, the previous season's harvest still remained unsold, the French wine makers of Bordeaux having being the only real market for these farmers.

At the beginning of 1941 *The Economist* reported Portugal was suffering a glut of produce from its colonies due to the British naval blockade of occupied Europe, the traditional markets for this trade. About twenty-two thousand tons, estimated at a total value of £1.5 million worth of oilseeds, coffee, cocoa and fibres were sitting in warehouses near the Lisbon wharves for want of buyers. These were matched by mounting inventories in São Tomé and Príncipe, Angola and Mozambique. These colonies had begun a few shipments to the United States, but these were described as a drop in the bucket. Producers in these colonies were penalised by the fact the British would allow through their naval blockade only enough produce for consumption within Spain or Portugal, anxious that nothing be available for re-sale to Germany.

## IV

India's big windfall came for its textile sector. Because the previous large shipments of textiles from Japanese factories were no longer available on the local market (and nor were British textiles), India's factories had the domestic consumer to themselves; and there were plenty of other markets which needed products that had formerly been bought from Japanese mills. On top of this, cloth demand for the military quadrupled. By 1943, cotton manufactures had quintupled in India and prices were soaring. And now the Indian mills had an export market, the Middle East, where prices soon rose to more than five times what they were in the Indian domestic market. A contemporary account conjectured that the profits of the mill owners 'should run into billions'. Pre-war, Indian cotton was priced at less than half a rupee per yard, yet in July 1944 *The Times of India* was reporting the cheapest woman's dress in the Middle East cost between Rs50 and Rs100 per yard. The Indian government then imposed maximum margins on export piece goods with the announcement stating the ceiling on margins was 'intended to enable consuming countries to obtain their supplies at reasonable prices'. Earlier in the year, the government's nine textile mills—seven of them in Bombay—were placed under central supervision, a measure seen by *The Manchester Guardian* in May 1944 as an attempt to drive out speculation and illegal trading in textiles.

However, it was not exactly a boom time for India's pig iron and steel producers, due largely to the shortages of coal. The mining companies suffered from a shortage of workers and the increased profits for coal were passed on only in part through wages.

Between 1918 and 1939 the overall number of factories in India had risen from just under 3,000 to 10,000. Of those, 1,700 were classed as large-scale plants. It was something, although not quite that impressive a number considering the size of India and its huge population. The industrial base was largely tilted toward processing of agricultural produce such as cotton, tea, just and sugar. It helped, too, that a substantial railway network had been constructed. Figures produced in the Indian Legislative Assembly in April 1943 showed that, for the three years to 1942, defence-related spending in India had amounted to £183 millions. In fiscal 1942-43, Britain had sent £45 million worth of war supplies to India. Britain also paid the cost of Indian naval defence apart from the £100,000 a year local contribution.

India's war industries were intended to supply not just weapons and ammunition for the defence of India itself, but also to take the pressure off Britain to supply of materiel to the Middle East front. The government of India had to begin from a standing start as the sub-continent was not geared for military production. As The *Times of India* reported at the end of 1940, 'in Bombay alone engineers have shown commendable ingenuity in producing, with old plant, precise and elaborate pieces of armament which a year ago experts would have sworn could not possibly be manufactured in India'. When war broke out, India was hampered by lack of machinery and trained heavy industrial workers. It is further evidence of Britain's underlying capacity that it was able to ship machine tools and industrial plant to India. Indian steel works were also faced with the challenge of lifting the grade of their product.

By 1942 the country's ordnance factories and railway work-shops were producing 700 different types of munitions. It has been

estimated that about fifty per cent of the national budget was for
the war effort. There were fifty-four plants making machine tools.
Artillery shell production had increased twenty-four-fold compared
to before the war, while the output of guns was up eight-fold. Factories were starting to build armoured vehicles (India had imported
all its armoured vehicles before this). Shipyards in India were launching new minesweepers, sub-chasers and other craft for the Royal
Indian Navy. Assembly lines were also putting together American
trainer and fighter aircraft: Hindustan Aircraft in Bangalore was producing Curtiss Hawk fighters and Harlow trainers—in both cases
only the engines were imported—along with a small number of
the American single-engine Vultee attack bomber and an Indian-
designed glider. India was also seen as one solution to the need to
replace all the British merchant ships sunk: in 1941 the government
gave land at Visakhapatnam on the eastern coast to the Scindia Steam
Navigation Company for a shipyard; however the first merchant ship
was not launched until 1948.

But certainly the ramp-up was impressive. By 1943 Indian
factories were turning out 50,000 miles of copper wiring a year
for telephone and telegraph use. The cotton textile factories were
able to produce some 138 different articles needed by the armed
forces. Woollen mills churned out vast numbers of army blankets, the
boot factories three million pairs a year. A canned food industry was
started from scratch.

Indeed, India was a vital economic cog in the Allied war machine.
The great fear after Japan entered the war was that India would
fall into the hands of the Axis and give it a chance of defeating
Britain and the Allies. Apart from the use of the Indian Army in
several theatres, including taking control of Iran (as well as in Eritrea,
Abyssinia, the Sudan, Somaliland, the Western Desert, Libya, Iraq,
Syria, Lebanon and Aden), the territory's other capabilities were
harnessed to the Allied effort. Indian engineers and skilled workers
were used to upgrade the Trans-Iranian railway to improve transport
links with the Soviet Union for the shipment of aid to the Russians.

India had ample supplies of iron ore (the Tata steel works at
Jamshedpur was the largest in the empire); it was the world's second

largest supplier of oil seeds, produced seven million bales a year of cotton, grew more tobacco than the United States, dominated the world market in jute, and mined one-third of the world's supply of manganese. India was the world's largest sugar producer during the war.

## V

Things had been looking up in rural Canada as the war approached. The 1939 crop had been better than most, so purchasing power was rising. Add to that the increasing military spending as Ottawa geared up for war; consequently, unemployed was falling and retail trade turnover improving. The mining industry had a great 1939. The gold price held and base metal miners found the munitions industry a ready buyer. War orders flowed through to the iron and steel plants and machinery suppliers. Textile mills hummed, agricultural machinery sales rose. On the month the European war began, the Canadian parliament levied increased incomes to pay for the military spending.

Canada was important for food production. In 1944 the country's farmers produced about 1.13 million tonnes of meat and milk production soared. This meant Canada could supply military food needs, maintain exports to Britain, but at the same time have plentiful supplies to meet domestic civilian demand.

There was also a huge war industry is in process of development at the beginning of 1941. In January that year Canadian plants had order books totalling C$849 millions for war-related equipment, more than half of that for the Canadian forces but C$309 millions from British buyers. More than 400 army trucks were rolling off assembly lines each day, 300 aircraft airframes were finished each month (and shipped to Britain) and tank production was about to begin. Eight factories were devoted to producing shells, another nineteen producing shell components. The plan was to get production up to two million shells a month. Bren guns were already in production by the start of 1941 and it would not be too long before various sizes of heavy artillery guns were also being shipped across the Atlantic. The next step would be into building warships, the British magazine *The Spectator* reported.

In Australia drought had hit the agricultural sector. Wheat exports had been disappointing; the poor performance in the rural sector led to a dip in domestic demand generally, which had flowed through to rising unemployment among manufacturing workers. The wool clip was smaller in 1939, too. The slowing of the Australian economy just as it was facing war was shown no more starkly by the import figure; at £99.38 million, it was down £12.34 million on 1937-38.

By 1944 the wheat acreage in Australia had been reduced by over one-third since 1939; hence, the wheat crop, while varying a great deal from year to year, has been smaller on the average than before the war. The 1944 crop, in consequence of a serious drought, turned out to be a complete failure; indeed, Norman Martin, the minister of agriculture in the state government of Victoria, went public with a concern in 1944 that the following year the country might have to actually import wheat to meet the needs of the population and United States forces in the Southwest Pacific that were dependent on food shipments from Australia for nearly three-quarters of what they required.

As the League of Nations reported in its 1945 summary of the war economy, the effect of the 1944 drought on the livestock industry was particularly marked in the case of sheep, and was reflected in the wool clip of the 1944-45 season. 'Millions of sheep perished and, although the cattle population as a whole suffered less, it is the consensus of informed opinion that the peak of Australia's beef production has been passed, and that output will decline until herds throughout the southern part of the continent can be restored. It is expected that this will take six to eight years to accomplish,' the report noted. The drought upset Australia's food production plans and, as a complication, added the factor a of great shortage of animal fodder.

Industry in Australia was quite a different matter. Manufacturing was harnessed to the war effort, meaning that domestic consumption of goods was forced to contract to the tune of twenty per cent; in fact, by the latter years of the war, the labour force involved in making consumer goods was less than a third what it had been pre-war. In September 1940 the Queensland state government placed its first order for the manufacture of shells and grenade casings with the Ipswich railway workshops. By 1941, the Toowoomba Foundry and the Ipswich workshops were producing machine tools for the armaments industry, while engineering companies were gearing up to build naval corvettes; by 1944, fifty-six had been launched across eight shipyards. But it was too slow getting started. When Japan entered the war, the Royal Australian Navy had just three of the Bathurst-class corvettes in service, but then production accelerated until, with the turning of the tide in the Pacific, it was decided to switch construction work to amphibious landing craft.

On 24 September 1941, *The Economist* described what it termed the 'Australian Industrial Revolution'. In January that year the first Bren gun was completed in an Australian plant, and tank production was gearing up. A version of the Bristol Beaufort bomber, adapted for conditions in the Pacific, was about roll off the assembly line, and smaller trainer aircraft were already in production with some shipped to India.

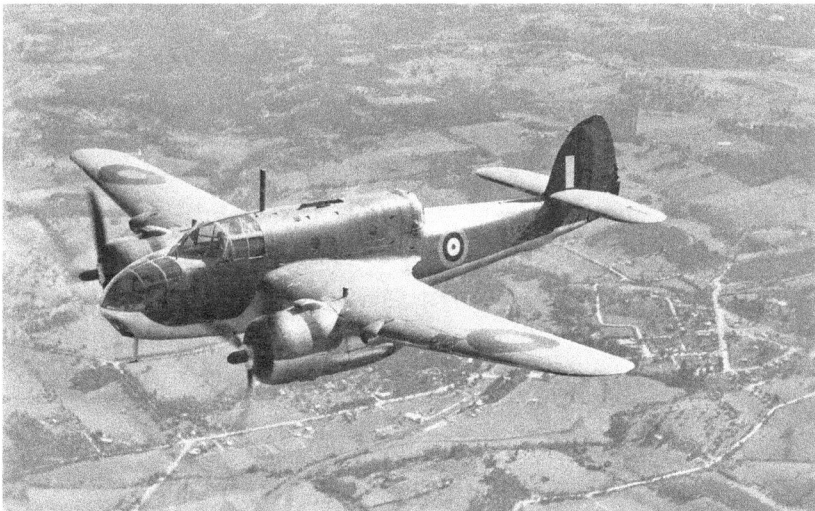

*Istol Beaufort L-4448, one of the two prototypes for Australian production. The aircraft is being flown by T. R. ('Tommy') Young, an engineer engaged by the Department of Aircraft Production to run production tests on Australian-built Beauforts.* Australian War Memorial.

Private investment, particularly in building construction, dried up. Rationing was imposed—from food to clothing to petrol—so there would be plenty to send to American forces fighting in the Pacific and stationed in Australia. Coal was rationed, too, as labour shortages and then labour disputes hit the industry

By 1943 munitions production at home was more than sufficient to meet campaign needs in the region, and it was possible to transfer some workers to shipbuilding and making much needed new agricultural machinery.

The loss of Nauru to the Japanese interrupted fertiliser supplies, and this was felt particularly in New Zealand. This exacerbated the food problem even with rationing of meat and butter (along with sugar, tea, clothing and petrol rationing).

In late 1938, across the Tasman Sea in New Zealand, the Labour government in Wellington had imposed severe import and exchange controls due to sharp drop in the country's sterling reserves. There was a shortage of imported goods which, typically, caused price rises. Labour, which had been elected with a huge majority in late 1935 and re-elected similarly in 1938 on the promise of a new social security system, had engaged on a large public works programme to stimulate employment, which resulted in farmers being unable to get sufficient workers. The foreign exchange situation was further complicated by a poor dairy season. *The Economist* in February 1940 noted that 'taxation is very high and business confidence is low'.

South Africa was suffering as a result of Germany no longer being an important export market for its wool and manganese. So far as wool was concerned, however, the British government had stepped in with increased orders. Gold exports remained the mainstay of the South African export sector.

## VI

On 21 January 1941, the Foreign Office in London issued a statement announcing Britain had undertaken to buy goods from the Belgian Congo, in return for which the Belgian authorities in exile would introduce 'trading with the enemy' legislation to apply to the Congo. The main item covered by the agreement was copper, of which the British contracted to buy 126,000 tons, that being a vital war need. By way of *quid pro quo*, Britain guaranteed to buy agricultural products, particularly palm oil and palm kernels, the export of which was vital to the Congolese economy. The announcement was not quite the full story: some agricultural products were bought, but not all as they were not actually required. Nevertheless, it was agreed that all exports from the Belgian Congo would enter Britain under tariff arrangements no less favourable than those applying to goods from Britain's colonies. But the Belgian authorities did have to make a severe financial concession: after deducting the cost of running the Congo, they had to hand over to Britain all of the gold output of their territory, plus the colony's foreign exchange reserves, in return for payment in sterling. This concession was as significant for the British war effort as was the physical copper shipments; it gave London much needed gold and foreign exchange with which to pay for the war effort.

Initially, the British were far more selective about what they wanted from the Congo. Soon after Belgium was occupied by the Germans, it was made clear from London that Britain wanted minerals, especially copper, but not the Belgian colony's palm oil, cotton or coffee. Both the Colonial Office and the India Office were concerned about the impact of Congolese exports to Britain vying with production from their own territories. But eventually it was decided in London that they could not risk the loss of Congolese support or alienate plantation owners and African peasants, and so some additional exports were acquired.

The Belgians in Leopoldville (now Kinshasa), unlike the majority of the French colonials, did not hesitate too long before swinging in behind the Allies and bringing with them the resources of the Belgian Congo. They had little choice: in 1939, eighty-five per cent of the Congo's exports had gone to Belgium, and from mid-1940 another market was needed desperately. The Belgian king had surrendered the army to Germany against the wishes of the political leaders, who wanted to fight on from France. The king ordered the governor-general in Leopoldville, Pierre Ryckmans, to observe strict neutrality, but that was ignored. It was not just a political matter for Ryckmans: the British blockade had disrupted the Congo's export business and he saw the advantages of selling to the British.

The Congo authorities set up a concentration camp near Elizabethville (now Lubumbashi) in which some two hundred German and Italians had been incarcerated. As the *Chicago Tribune's* paper's roving foreign correspondent Maxwell M. Corpening informed the paper's readers on 14 March 1941 in a dispatch from Elizabethville, the Belgian Congo covered 902,000 square miles and was home to ten million Africans and 25,000 Europeans.

The Americans arrived soon after they were dragged into the war. By this time, too, Pan American Airways was crossing the Atlantic from Brazil. An American woman, Dudley Barinen, was to work for the Free French Information Service across the Congo River in Brazzaville, in French Equatorial Africa. Sitting in Leopoldville in July 1942, she created a pen portrait for readers of the *Christian Science Monitor*, noting that beyond the city limits there lay the great equatorial forest in which 'the natives wear feathers and hunt elephants'.

> Yet here in so-called darkest Africa I have just ordered my waiter to bring me a peanut butter sandwich. I am wearing the latest American cotton dress purchased at a shop down the block. At my table are some American army officers who had dinner at the Waldorf only a few nights ago, for the Pan American clipper service has put the skyscrapers of Miami within four days of the jungles of the Congo.

Turning to the needs of war, Barinen noted the Belgians were targeting 10,000 tons of tin being mined, up from 5,000 tons in 1941. There were also attempts being made to re-establish the largely abandoned rubber industry, production in the Congo always being from wild trees rather than plantations. Planting of rubber trees was under way as she wrote. Palm oil was a well established industry.

'Along with many innocent Americans, I arrived here lugging a suitcase full of soap, unaware that I was coming to the home of palm oil, basis ingredient of this product,' she noted.

Similar deals to those done with the Belgian Congo were put in place with General De Gaulle's Free French administration. Britain agreed to buy the total output by the French Cameroons of cocoa, palm kernels, palm oil, ground nuts and beniseed (a plant that produced high protein oil suited to use in food products) as well as almost all the Cameroon's coffee crop and much of its banana crop. The British got much needed food items while the farmers of the French Cameroons now had a guaranteed market. It was followed by a similar agreement covering French Equatorial Africa. The potential riches of the latter territory (now the Republic of Congo) had been realised by the British in 1940: apart from cocoa, nuts and food oils, French Equatorial Africa was seen as a valuable source of hides and skins, resins, manioc, and some hard fibres, along with wild rubber, a variety of tropical timbers. Limiting its value, on the other hand, was that France's equatorial colonial territory was the least developed part of its empire, with just 3.5 million people spread across its vastness. However, Brazzaville's loyalty was locked in when Britain agreed to make up for the loss of the subsidy from France which was essential to maintain the government, with London providing £200,000 and the agreement to buy all that French Equatorial Africa could export, a concession which it was hoped would be noted in those territories still loyal to Vichy.

Then there was the question of logistics. The virtual absence of pre-war inter-colonial trade in Africa meant that transport systems, railways where they existed, connected ports to their hinterlands but there was no infrastructure that allowed any significant land-based trade between various colonies. This meant that, in the case of Chad, there were no all-weather roads connecting that territory with British operations in either Nigeria or Sudan.

By 1941 colonies that had swung over to the Free French side were French India (population 293,000), French Equatorial Africa (population 3.5 million), St Pierre et Miquelon (population 3,916), and New Caledonia and French Polynesia (population 100,000). Vichy still controlled North Africa (and its sixteen million people), West Africa (fifteen million), French Somaliland (50,000), Martinique, Guadeloupe and French Guiana (496,000) and Madagascar (four millions). But Madagascar would soon be lost after a British-South African invasion, West Africa after military invasion, while the Caribbean possessions were being watched by the Americans. French Indochina and its 23.2 million people were, effectively, under Japanese control.

Nothing could be done by the British and the Dominions about the Vichy government on the French mainland, but those overseas colonies that declared allegiance to Marshall Philippe Petain were quite another matter. Whatever the state of Britain's colonies, as *The Manchester Guardian* commented in late 1940, was by contrast that the French had done so little to develop their territories. French India, which consisted of five enclaves along the east and west coasts of India, was dominated by Pondicherry (now Puducherry), the only French colony which boasted any manufacturing industry. Textiles were an important part of the economy (Rodian mill, for example, employed eight hundred workers) and the ports of Pondicherry and Karaikal were centres of bustling commerce and exporting. Beneath that bustle, though, French India was not a happy place in the late 1930s: there were rolling strikes in the textile mills provoked by appalling working conditions, leading to riots and shootings by police. Smuggling was also endemic as traders tried to avoid customs duties imposed in British India.

Apart from that, French India's significance before the war was mainly as a reminder of French aspirations in India two centuries earlier, and as a stopover for French officials travelling to and from Indochina. French India's enclaves in total covered little more than 500 square kilometres and were home to just over 300,000 people. Pondicherry was the largest, the home of the governor of French India and just over 170 km from Madras; then there was Karaikal further south; Yanam also on the west coast; Mahe, little more than a village on the east coast due west of Mysore; and Chandernagore located just 48 km from Calcutta. By 1949, just five years away from India reclaiming these enclaves, most of the business was being carried out in rupees, although over-printed franc notes from Indochina were also in circulation.

Given its dependence on British goodwill, French India in December 1940 declared it would support Britain and de Gaulle. Governor Louis Bonvin, in opening the French India legislature that month, declared his territory was 'ready to go to the limit of its resources in fighting side by side with the noble British', a somewhat different sentiment that prevailed in the other French colonies (with the exception of Chad). A delegate from de Gaulle's Free French movement was present in the legislature, according to a Reuters report. Almost a month later, the people of Pondicherry staged a hartal—a general strike—to protest their support for de Gaulle. The general's representative in Pondicherry had requested that, on New Year's Day 1941, all people should stay at home in the afternoon and pray for the liberation of France. Accordingly, all shops, hotels and restaurants in Pondicherry were closed between one and four in the afternoon of 1 January.

And, thus, after these protestations of loyalty to Free France and the French empire, life went back to pretty well normal in French India. There were a few changes. While British India retained the right of appeal from the Indian courts to the Privy Council in London, a similar right in respect to the de Gaulle administration in exile was impractical for French India. The de Gaulle administration gave the final decision to the Pondicherry Supreme Court, although any death sentences would be suspended and convicted killers could remain alive until the war ended, upon which time their appeals could once again be heard in France. A customs union was soon signed with British India, imposing the same duties at Pondicherry and Karaikal that applied at the other ports on the sub-continent. Once the French Indian colony of Pondicherry had thrown in its lot with the Free French, the British authorities in India did all they could to support the French in Pondicherry and the smaller French enclaves along the coast. The Government of India opened a post office in Pondicherry and, French India never having had air mail stamps, it was decided to issue Indian stamps until a new set of sixteen stamp denominations for the French territories could be printed in London. The Cross of Lorraine, the Free French symbol, was used as a cancelling mark on the stamps.

## VII

One Vichy outpost was just twelve nautical miles from part of the British commonwealth, the Dominion of Newfoundland, a territory that had long rejected union with Canada (although it finally succumbed in 1949). The French islands of St Pierre and Miquelon had been demilitarised under a 1763 Anglo-French treaty, so there was no threat of military action from the tiny territory. But that was only part of the story because a shortwave transmitter at St Pierre allowed communication with France (and the transmission of weather reports that would have been useful to German U-boats in the Atlantic), and there was a fleet of French deep-sea trawlers working the Grand Bank of Newfoundland which all had high-powered radio sets. In addition, there were the sub-Atlantic cables that connected the United States and Canada with Britain passing close to where the French fishing fleet operated. The fear was the radio facilities would be used to send information about convoys to the Germans, a fear exacerbated by the fact that it took more than a year to break the codes being used by the station at St Pierre.

*Time* magazine on 5 January 1942 described territory thus: '93 square miles of rock and its population of 4,321 gripped by poverty'. The islands in themselves were of no importance 'except as a great cod-fishing centre in the 1880s and as a lush rum-runners' rendezvous in Prohibition days'. With the invasion of France, the Canadians had tried to help the islands, offering to market the dried cod produced and guaranteeing the convertibility of the St Pierre franc. Moreover, Canada did not want to be forced to send troops to the islands and the only soldiers available to the Free French were black troops from French Equatorial Africa. Another factor weighed on minds in Ottawa, and that was the fear of offending French Canadians by invading St Pierre and Miquelon. There was a different attitude in Newfoundland: there the talk was of occupying the islands.

The tension started to rise. Canada banned all meat shipments to the islands in August 1941. This was followed by a decision in Ottawa that a consulate would be opened in St Pierre to monitor what was happening there. Alarm continued to grow when the Vichy government extended its anti-Semitic policies to the Caribbean territories of French Guiana and French Antilles as well as to St Pierre and Miquelon, along with ordering a census of Jews in these territories.

Then came action. Three Free French corvettes arrived at St Pierre on Christmas Eve, with Admiral Émile Muselier taking control of the territory. The American papers had the story overnight, although they reported there were four ships. 'Sailors and marines in full war gear leaped ashore and raced to take possession of strategic points, including the Governor's residence, the police station, the wireless and telegraph offices and customs,' reported *The New York Times* on 26 December, adding that the inhabitants greeted this with cries of 'Vive de Gaulle'. Muselier carried a list of Vichy agents who were promptly rounded up, including the stridently pro-Vichy governor Baron de Bournat. He also ordered a plebiscite to held the following day; the reporter aboard the admiral's ship said that he had been told the Free French would withdraw if the vote was pro-Vichy. That issue did not arise: there was overwhelming support for the islands to throw in their lot with de Gaulle.

Washington was quick to respond—and the Americans were not happy. Secretary of State Cordell Hull called on the Free French to withdraw. The concern in Washington, according to the correspondent of *The Los Angeles Times*, was that this was exactly the sort of action that could push the Vichy regime further into helping the Nazis. The US policy since the fall of France had been one of limited co-operation with individual Vichy colonies. But within two months, the affair was largely forgotten. There were bigger issues to deal with when it came to Vichy territory outside France.

# 6.

# The Axis loses in Latin America

HARRY S. TRUMAN IN 1942 headed the United States senate's Special Committee to Investigate the National Defence Program. At a hearing on 3 April that year, army and navy intelligence services furnished a list of names of those who had recently travelled to and from Latin America on the Italian airline LATI (Linee Aeree Transcontinentali Italiene). The passenger lists included Arnulf Fuhrman, Gauleiter for South American anti-semitic propaganda (who in 1940 had been arrested in Uruguay for trying to establish a fascist regime), Erich von Ribbentrop (nephew of the German foreign minister and Bund leader for Colombia), Colonel Rudolf Meissner of the Gestapo, several generals, Adolph Paez who was described as 'storm trooper officer to work through schools', various propaganda agents. Among the generals using the LATI service to Rio de Janeiro was Eberhardt Bondstedt, whom the Truman committee was told had been authorised by Berlin to act as military leader for Central America, his assistant being accredited to the German embassy in Panama City.

By the time of the Truman hearings, the LATI flights to Brazil were already a thing of the past: in the previous December, the last Italian airliner took off from Brazil, the service thereafter being cancelled because the Brazilians had made it clear they would no longer provide gasoline for the return journeys. In December 1942, LATI's man in Brazil, Count Edmondo di Robilant, was sentenced by a Rio de Janeiro court to fourteen years in gaol after being convicted of being an Axis spy. One of his aids received twenty years, three were give fourteen years and another eight years; one man, Giovanni Pianezola, was acquitted. The count had achieved some fame in Brazil in 1931, having crashed into the jungle and being found alive after eighteen days. Di Robilant was reported to have confessed to the Brazilians that he sent radio messages to Rome giving details of Allied shipping, including the visit by the *Queen Mary* (then a troopship) to Rio. Apparently di Robilant had cajoled an Italian migrant living just outside the city to keep the transmitter buried in an iron box at the latter's rabbit and guinea pig farm.

The LATI aircraft, a tri-motored Savoia Marchetti S83T converted bomber, took three days from Rio de Janeiro to Rome via the Cape Verde Islands, Rio de Oro (now Western Sahara) and Seville. Apart from Axis officials, the flights also carried important cargoes to South America including books, maps, chemicals, films, aviation equipment, photographic materials and medals; on the return flights, the Italian aircrews flew mica, semi-precious stones, platinum and diamonds. The report to the Truman committee said ninety per cent of Europe-bound cargoes were essential war materials. Even though it was an Italian airline, the Axis passengers were mainly Germans and the cargoes films and books in German.

The testimony to the Truman committee was not news to the U.S. government. As early as June 1940, the American ambassador in Montevideo, Edwin C. Wilson (who would show equal perspicacity as Ambassador to Turkey after the war) cabled to Roosevelt the advice that several countries in South America were in danger of falling under Nazi influence. The Federal Bureau of Investigation, helped by the British agents in the region, soon came to the conclusion that LATI was the conduit for German and Italian couriers and agents; the problem was, before the U.S. was at war, that a number of influential Brazilians were supporters of the Italian airline.

Both Germany and Japan had been eyeing the potential of Latin America, especially as a source of raw materials; Japan took about one-third of Peru's cotton crop for example, and, as noted earlier, the Axis countries would view the region as a potential source of metals vital to their war effort.

In Washington, meanwhile, there had been mounting apprehension about German and Japanese intentions to America's south. By 1941, the U.S. was concerned that islands of the Caribbean could be used as bases for an invasion of the American mainland (the northern route invasion route having already being secured with the occupation of Greenland and Iceland). Moreover, the U.S. clearly did not want to have to divert too many forces to defend itself from attack from the Caribbean or South America. And then, of course, there was the Panama Canal.

South American civil aviation had, in the years before the war, been heavily influenced by German interests. German companies operated three airlines directly: Condor in Brazil, Lufthansa Peru and SEDTA (Sociedad Ecuatoriana de Transportes Aereos) owned by Lufthansa in Ecuador—and had nationals in senior positions at Bolivian airline LAB (Lloyd Aereo Boliviano) and Brazil's Varig, while the Italians influenced the running of Corporacion in Argentina. Three other airlines used German aircraft: CAUSA in Uruguay, Aeroposta in Argentina and VASP in Brazil. In all, forty German and Italian-made aircraft were flying South America's air corridors. That they were heavily subsidised from Europe was obvious. The United States carriers in South America stuck to the most profitable, high density routes—along one coast through Colombia, Ecuador and Peru to terminate at Santiago, Chile, and along the other through British Guiana, Natal in Brazil, to Rio de Janeiro then terminating at Buenos Aires. By contrast, the routes flown by German controlled carriers offered far less potential profit. One service run by Condor left Sao Paulo and landed at Corumba, a Brazilian town not far from the Bolivian border; then the aircraft flew on to Santa Cruz de la Sierra in Bolivia where passengers transferred to Lloyd Aereo Boliviano for the leg to La Paz. Having disembarked passengers for the La Paz flight, the Condor airliner flew on to the northern Chilean city of Arica, just eighteen miles south of the Peru border. One theory about choice of routes was that the Germans saw them as more of a diplomatic move —offering government officials frequent services to the interiors of their own countries—rather than a plan for business success. The Ecuador operation also carried mail at below cost rates.

Condor before the war provided a connection to Lufthansa's trans-Atlantic services. In 1930, the Graf Zeppelin airship had made its first landing at Rio de Janeiro.

Even though the German airlines were cut off from their home country, they were expanding services in Latin America from 1939 and had no apparent problems acquiring finance, supplies and spare parts; Condor held large inventories of spare parts and these were made available to all the German carriers. In March and May 1941, German ships managed to run the British blockade and arrive in Rio de Janeiro with aircraft (two Junkers Ju-52 three-engine airliners), spare engines and twenty tons of spare parts.

The concern in Washington was that the Germans might build an airbase in South America from which they could bomb the Panama Canal. Pressure was applied to Ecuador and it was arranged for a U.S. air company to take over SEDTA's operations. The pretext seized on by the government of President Carlos Arroya de Rio was to declare SEDTA's fuelling practices unsafe; when supplies of aviation fuel were running low, the airline mixed in gasoline to make the fuel go further. SEDTA was sent packing and the Americans snapped up the work in Ecuador. Across the southern border, the Peruvian government in April 1941 announced it was closing Lufthansa Peru and seizing its aircraft.

There was also what *The Economist* called 'a remarkable enterprise'. New Zealander Lowell Yerex had founded Taca (Transportes Aeros Centro-Americana) serving the small Central American republics of Honduras, El Salvador, Guatemala, Costa Rica and Nicaragua. His enterprise headed off any attempt by the German airlines to penetrate north of the Panama Canal.

## II

Even before the United States joined the Second World War, the Germans seemed as intent to turn Latin Americans against Washington as they were to sully Britain's name and reputation. As *The New York Times* informed its readers just a month after Britain and Germany had gone to war, the Nazi regime had long been supplying a free news service to South American newspapers from its Transocean News Agency based in Panama. Axis-oriented newspapers were widely available—in Buenos Aires, for example, newsstands stocked the German-language *Deutsche la Plata Zeitung,* the Italian-language *Giornale d'Italia* and a Spanish paper, *Diario Espanol,* published by Franco sympathisers. At this stage, Germany was also setting up a new propaganda headquarters in Montevideo and was planning to distribute daily news bulletins totalling between 20,000 and 30,000 words, Uruguay being chosen as it had much less onerous censorship laws than other South American states. In Panama, radio station La Voz del Pueblo, broadcast German-supplied news bulletins several times a day.

The Germans used the postal services to target influential Latin Americans with books and pamphlets. In Lima and other capitals there was widely circulated a children's Spanish-language edition of the story of Hitler's life. In July 1940, postal authorities in the Misiones territory of northern Argentina estimated that, by weight, ninety-five per cent all of the mail handled was German propaganda. But that was only the surface expression of German influence in the region. There were some 1,400 German schools in Brazil. And trade was done on a bartering basis: manufactured goods from Germany in the years leading up to the war were paid for with oil or minerals, an exchange system that was ended only by the British naval blockade. (As late as December 1939 some thirty-four German merchantmen were still locked up in Mexican, Argentine, Chilean, Brazilian and Uruguayan ports; the *Nienburg* and *Anatolia,* for example, were waiting with their wheat needed urgently in Germany.) In early 1939 Bolivia sought help from Germany in operating the oil fields that had been confiscated from Standard Oil, and Berlin had offered large credits to La Paz in return for all the oil to be produced. A German team, led by a Colonel Bruno Metlitzky, was also sent to Bolivia to work out a barter arrangement with Bolivia supplying key raw materials in return for German weaponry. The colonel was representing the Siemens group.

As war with Japan—and, therefore, Germany—neared in July 1941, the Americans felt they were getting ambivalent messages from La Paz. The Germans still had strong connections inside the Bolivian army; during the 1932-35 Chaco war with Paraguay, Bolivian forces had been helped by German instructors. Yet the civilian government was much more amenable to the Allies, as evidenced by the takeover of Lloyd Aereo Boliviana and the transfer of its routes to Pan American Airways.

Germany also offered to build a 350 km highway from the Paraguay capital of Asuncion to the Iguazu Falls where the Argentine and Brazilian borders met. Germany proposed to also encourage various industries, textile mills, tobacco factories among them, which could use that road to export their products. Paraguay had previously tried to interest American companies to build the highway as a toll road but without success.

Paraguay was fertile ground for the Nazis. Indeed, the landlocked country stayed sympathetic to the German war effort until 1945. Higinio Morínigo became president in September 1940 after the death of President José Félix Estigarribia in an aircraft crash. The U.S. administration tried to woo Paraguay away from its pro-Axis leanings by offering financial assistance and technical aid. But this failed and it was not until the Japanese bombing of Pearl Harbor that Washington was able to force Morínigo to break diplomatic relations with the three Axis powers (but he stalled on declaring war on Germany until February 1945 when the outcome of the war was seen by the Paraguayans as beyond doubt). While the government in Asuncion played along with the Americans on the surface, they remained committed supporters of Washington's enemies beneath that surface; large numbers of Paraguayans had been converted to the Nazi cause by German agents and there were German schools, hospitals, youth groups that openly displayed the swastika and pictures of Adolf Hitler. The newspaper, *El País*, was overtly pro-German in its editorial policy. One of the more extraordinary acts of admiration was that by Paraguay's police chief who gave his son the personal names of Adolfo Hirohito. This man's police cadets wore uniforms with swastikas and Italian insignia.

In Bolivia, Germán Busch, the president until just before the outbreak of the European war—and whose administration was forging many commercial deals with Germany—was the son of a German doctor (and Italian mother). His Minister of Mines and Petroleum, Dionisio Foianini, was not only the son of an Italian father, but studied pharmacy in Italy and became an admirer of Mussolini's. So pro-fascist was Busch that in May 1939 the Italian newspapers reflected the hope that Italy had in Bolivia the first South American recruit for the Anti-Comintern Pact, the alliance forged between Germany and Japan in 1936 and joined by Italy the following year and later by Hungary, Manchukuo, and Spain; as German domination of Europe increased, several others—including Bulgaria, Romania, Denmark and Finland—also signed up. As it turned out, it was a false hope in Bolivia's case.

Much of the Axis activity is reminiscent of the activities portrayed by 1940s American movies about Nazi sabotage. Someone who seemed to personify the stereotypic Nazi was Eric Cerjack-Boyna, described in one account as 'a mysterious person of Hungarian origin'. He was ostensibly in the import business in Panama City, but was regarded as the paymaster for those conducting Nazi subversion in the country. Although not a member of diplomatic service, he was accorded diplomatic status by the Panamanian government.

He pops up occasionally in news reports of the time. One, filed by the Associated Press in August 1940, has him appearing in an American court located in the Canal Zone where 52-year-old Dr Emil Wolff—and one immediately pictures someone who could have been portrayed in one of those American movies by Peter Lorre—was fined $2,000 and give a three-month suspended sentence in the U.S. penitentiary after pleading guilty to being a foreign agent. Wolff had sailed from San Francisco with a trunk containing diplomatic documents from the German consul in that city and admitted to be carrying them to the German consulate in Valparaiso, Chile. Cerjack-Boyna told the court that the fine would be paid immediately and Wolff would leave for Chile.

In July 1941, five months before the U.S. would enter the Second World War, the State Department issued a long list of persons and companies in Latin America who were being blacklisted. Cerjack-Boyna, whose address was given as 17 Sixth Street, Panama City, was one of those named. (The extent of the blacklisting included such names as the Argentine branches of Mercedez Benz, Agfa and Bayer and, in Panama, Italy's Lloyd-Triestino shipping line.) *New York Times* correspondent Benjamin Welles wrote of Cerjack-Boyna in August 1941 that 'although his name recently appeared on the American blacklist, he is said to be on terms of easy affability with many prominent Panamanians'. In fact, the report continued, German charge d'affaires Hans von Winter was regarded by the Panamanians as subordinate to Cerjack-Boyna. The newspaper said about 900 Germans were living in Panama and almost all of them had joined the local Nazi Party.

Some of the Central American rulers displayed a particular predilection for the Axis. Maximiliano Hernández Martínez, who ran El Salvador from 1931 until 1944, was one such. In early 1940, he decreed it to be a crime to express praise for the Allied cause. That same year, upon receipt of the news of Italy's entry into the war and invasion of southern France, three hundred Blackshirts paraded in support through the streets of San Salvador.

In 1938, a small group of Italian aviators was based in San Salvador after the country took delivery of five two-seater Caproni AP-1 fighter-bombers from the Breda company in Italy (the El Salvadorans having exchanged $200,000 worth of coffee for the aircraft). This was not the first tie-up between the military of this small country and the Axis. In 1936 El Salvadoran officers had begun living and training in both Germany and Italy. However, economic reality was biting by late 1940; the inability of Germany and Italy to ship their exports to El Salvador was starting to cause several shortages of certain goods in the Central American state. Martinez faced reality and in October began praising the Allied cause. He found that Washington was, consequently, receptive and happy to supply rifles to his soldiers.

### III

Brazil had a substantial trade with Nazi Germany before the war. In 1938, Brazil was Germany's ninth largest export market. That year Germany imported from Brazil a total of 466,364 bales of raw cotton, 91.8 million kilograms of coffee (forty-one per cent of Germany's imports) and 10,600 tons of cacao. In that year, too, fourteen per cent of Germany's tobacco was imported from Brazil; more than three-quarters of the country's wild rubber produced in 1938 was shipped to the Reich.

With Germany the Brazilians did not need to eat into their scant supplies of hard currency; instead, they bought German manufactures with what were known as Aski marks (for Auslander Sonder Konto fur Inlandszahlungen, meaning 'foreign special accounts for domestic payments'). These marks were released by Berlin only for approved transactions. The Germans had bought cotton, wool and fruit with Aski marks, and in turn the Brazilians used those for purchases from German industries.

Brazil continued to trade with Germany until the outbreak of war, at which time the British naval blockade closed off the shipping routes to the Nazis. Then the country made its position clear: it not only allowed U.S. forces to base aircraft (at Natal) and naval ships (at Recife), but sent troops and fighter planes to fight with the Allies in Italy. Brazil declared war on Germany in August 1942 after the sinking by U-boats of several Brazilian merchantmen inflamed public opinion and demonstrators called on the government to join the fight against Germany.

Brazil prospered as a result. As American historian Frank D. McCann points out, its alliance with the Allies led to a large hard currency reserve situation for the first time in more than a decade. Its 1942 trade surplus amounted to $148 million and by the end of the year Brazil had increased its gold reserves to the value of $121 million, triple the figure for 1939. Textile factories were kept busy with foreign orders, especially from Argentina and South Africa.

The trade boom enabled Brazil to start to emerge as an independent economic power, and to loosen the control of foreign investors. Before the war, as McCann wrote,

> foreigners controlled street car lines, electric power, coal and oil importation, much of the flour milling, all of cement production, many of the tugs and barges in Rio's harbour, and telegraphic communications with the rest of the world. A British company had owned the sewers of the older parts of Rio since 1857. Many of the movie theatres in big cities were owned by Paramount, RKO, and Twentieth-Century Fox, who actively discouraged development of the national cinema industry.

## IV

German investment in Latin America grew steadily during the inter-war years. J. Fred Rippy, then professor of American history at the University of Chicago, did an estimate of that flow of German money. In Argentina, he believed the investment total in 1918 may have been $250 million, then in 1924 it reached $375 million and by 1939 it was around $540 million. What is even more striking is that the number of German nationals living in Argentina doubled between 1918 and 1939; to them you had to add large numbers of Swiss, Austrians and Germans who fled Russia after the communists came to power there. It was estimated that, in 1931, there were 725,000 people of German descent living in Argentina, around 50,000 of whom still held German nationality. In the early years of the war, Buenos Aires was home to branches of many German firms: there were the shipping companies Hamburg-American and North German Lloyd, Banco Alemán Transátlantico and Banco Germánico,

Siemans, Thyssen and Mannesmann, Bayer, as well as insurance, construction, real estate, machinery and railway companies.

An Argentine writer, Luis V. Sommi, in 1943 had published a study of German financial tentacles in his country. He identified the electric trust AEG as the largest German enterprise; it was operating telephone companies, the Argentina Electric Company, provided insurance services and ran boarding houses, not to mention owning large rural (cattle, wheat and corn) enterprises. Siemens-Schukert, though its International Telephone Company, controlled all the telephone and telegraph services to and from most of the provincial areas of the country. Lahusen y Cia was an important player in the wool export trade. Sommi also found that Spanish help was sought to avert any Allied moves to try and confiscate German companies; the Spanish put money into several enterprises, enough for them to qualify as neutral-owned operations. Thus the German Overseas Electric Company in Argentina was transformed into the Hispano-American Electric Company.

## V

Japan, by comparison, had modest connections in Latin America. Until the 1930s trade between Japan and the region was insignificant and diplomatic relations were not given great importance by either side; indeed, it was not until 1935 that anyone was accredited by Tokyo to Costa Rica, El Salvador, Honduras, Guatemala or Nicaragua. It was only when Japan began to prepare for war that greater efforts were made in trade; in the mid-1930s, purchases were made in Mexico of petroleum, scrap iron and mercury. There was talk by 1939 of a pipeline being laid to Mexico's west coast to load Japanese oil tankers but nothing came of the plan.

There had been, however, one slightly bizarre incident in El Salvador's relationship with Japan. In March 1934, due to a slip-up at the foreign ministry in San Salvador, some official had triggered the country's official recognition of the Japanese puppet regime in northern China, Manchukuo. As reported by *Time,* the day before

Henry Pu Yi—who as a child had been the last emperor of the Qing dynasty—was to become Emperor K'ang Te of Manchukuo, his diplomats in Hsinking sent an announcement to foreign ministries around the world proclaiming the enthronement. League of Nation members, of which El Salvador was one, should have observed the league's resolution of non-recognition. Instead, San Salvador sent a cable offering 'fervent congratulations on this happy occasion'. That was enough for Hsinking to claim recognition by San Salvador. But there was also a bizarre follow-up: in 1938, *The New York Times* reported that K'ang Te had, via the Manchukuo embassy in Tokyo, delivered decorations to President Maximiliano Hernández Martinez, foreign minister Miguel Araujo and the El Salvador consul in Tokyo. The El Salvador government did not walk back from their 1934 act: in 1939, they appointed one Wang Ching Shan as their honorary consul in Hsinking.

While the main concern in Washington about Japanese in Latin American focused on the comparatively large numbers in Brazil, the ethnic Japanese had settled widely. Just under eight hundred had migrated to Peru in the last years of the nineteenth century, and some had soon moved into Bolivia to find work. Eventually a few trickled into La Paz and by 1922 there were enough of them to found the Japanese Association in the Bolivian capital. When the Pacific war began, the twenty-nine Japanese living in La Paz were rounded up and interned in U.S. camps.

There were American fears that Tokyo had its eyes on colonising Peru. The suspicions was that Tokyo had its eyes on controlling Peruvian cotton in order to keep Japan's piece goods mills supplied. There were by the beginning of 1942 an estimated 25,000 Japanese living in Peru, although it was noted by *The Christian Science Monitor* that many Japanese deliberately avoided the census count, and there was a school of thought their numbers were closer to 50,000. In their communities, the Japanese ran their own primary and secondary schools, teaching in their mother tongue. It was reported that several Japanese who were members of the military reserve at home had bought land in Peru.

Japanese policy changes had seen more overtures being made to the region by the mid-1930s. In 1938 Tokyo had made approaches to Cuba seeking a new commercial treaty. One area that was of potential interest to Japan—which sold mainly textiles to Cuba—was the island's iron ore, manganese and copper, most of which was then exported to the United States.

Japan also saw Latin America as a key source of raw materials. In 1941 it was reported that Japan had placed orders for about one-half of Peru's cotton crop, was purchasing molybdenum and would have bought copper as well had not the U.S. already ensured that Peru's supplies of the red metal went to them. In fact, Washington bought 36,000 tons of Peruvian sugar in 1941 even though it was itself a substantial sugar producer, just to keep it out of Japanese hands. Meanwhile, the Peruvians required sales to Japan to be paid in dollars, not yen. Nevertheless, Tokyo was able just months before going to war in the Pacific to buy strategic metals from Peru, including antimony, tungsten and vanadium.

By September 1940 Washington had negotiated a new trade deal with Peru—thus gaining first priority to that country's output of copper, vanadium and lead—and by this time was also in the process of tying up metals produced in Argentina, Bolivia and Chile. Mexico and Brazil had already pledged their strategic metal exports to the United States. Three months before the attack on Pearl Harbor Washington was already waging full-out economic warfare in Latin America.

## VI

While the two belligerent sides vied for advantage in the region, the countries of Central and South America sailed through the war relatively unscathed. As the League of Nations explained in its 1945 summary of the economic consequences of the war, the Latin American states increased their exports during the war; but because the Allies were exporting greatly reduced amounts, or not exporting some things at all, the imports into Latin America dropped sharply.

This had two effects: one, foreign exchange reserves (including gold) increased for most Latin American states and, two, it encouraged development of import substitution manufacturing industries in these countries—however, the latter was limited by the fact that plant and machinery were not readily available, so much existing equipment had its uses changed or utilisation increased. The Argentinians found they could not get parts for mechanical equipment, while cotton mills in Brazil and the railways in Mexico had to keep running with equipment that was getting old and deteriorating. (The Mexican railway system had been put under pressure in 1942 when U-boat activity in the Caribbean forced cargo off ships and on to land transport.) The shortage of imported fuels forced development of hydro-electric plants in Brazil, Chile and Uruguay. And the unavailability of enough imported steel led to a surge in steel-making in Brazil, Mexico and Uruguay.

The downturn in imports also posed another problem: much government revenue in the region was derived from import duties, so direct taxes had to be introduced or increased to make up the difference. Venezuela introduced income taxes for the first time, while Chile tripled the rates of income tax three-fold over a four year period.

In terms of industrialisation, Brazil was the greatest beneficiary. The number of industrial workers increased by around fifty per cent by 1943, although at about 1.5 million it was still a fraction of the per capita proportion of the population common in Europe. More importantly, perhaps, the exports of manufacture goods rose 3,562 per cent as Brazil shipped tyres, textiles, iron and steel, machine tools and china and glassware to its regional neighbours. Coal production doubled.

# 7.

# Italy—the weakest link

## Italy's Empire 1940

|  | Area (sq. miles) | Population |
|---|---|---|
| Eritrea | 90,000 | 1,500,000 |
| Italian Somaliland | 271,000 | 1,150,000 |
| Abyssinia (Ethiopia) | 305,000 | 9,450,000 |
| Libya (inc. Libyan Sahara) | 680,000 | 888,400 |

SELF-SUFFICIENCY THROUGH ITS empire—that was Italy's grand plan as it entered the Second World War at the side of Hitler's Germany. That self-sufficiency was never achieved and, really, the Italians never really tried hard enough. As historian R. J. B. Bosworth notes, no effort was made to extract known gas reserves in Italy itself; in the thirty years Italy controlled Libya, that colony's vast oil reserves lay undisturbed. 'Italian economists and geologists, learned in the classics, preferred instead to fantasise about "restoring" Libya to the position it had once held in imperial Roman grain production. Until the fall of Fascism, the colony's greatest mineral export was salt,' he writes.

At the first meeting of the grandly named General Councils of the Corporate Councils of Italian Africa in January 1940, the

Minister for Italian Africa, General Attilio Teruzzi, set out a resources production plan. This would begin with ensuring adequate food supplies for the populations in the Italian possessions, but production would also have to meet the needs of war and supply of those items of which Italy itself was deficient. Teruzzi specified cotton, oilseeds and metals. The 'great final goal' was to open these lands to Italian settlers.

The Italians saw Abyssinia, which they had invaded in 1935, as the jewel in what was otherwise a rather useless colonial empire. By 1939, Teruzzi was claiming that 200,000 Italians had moved to the new colony, including 2,000 farmers on plots of land ranging up to 120 acres. The Italian authorities were encouraging coffee and oil seed cultivation, were building a new textile plant and were planning to develop gold mining in the western part of the country.

Some, though, made the comparison between the empire with which Italy had joined the war—Eritrea, Italian Somaliland, Libya and Abyssinia—with that of the Roman Empire at the time of Julius Caesar, which was bounded by the Euphrates and Armenia in the east, by Abyssinia to the south, the Danube to the north and by the Atlantic to the west; present day Morocco, Algeria and Egypt had also been subject to rule of the emperor in Rome at varying times.

Italy got the leftovers of the scramble for Africa. It had acquired Eritrea and Somaliland in 1890 and 1905 respectively, and then in 1911 attacked Libya when Rome judged the Turks to be too weak to hold the territory. Italian troops landed in Tripolitania, the north-western part of what is now modern Libya and which includes Tripoli; but while Rome wanted colonial conquest, no real thought had gone into what they would do with their new territory once they got it.

However, the Italians had convinced themselves that a combination of inept Turkish rule and indolent Arab farmers had held back the territory, the implication being that the Italians could get rid of the Turks and put in farmers who knew what they were doing.

But it was not until the Fascists under Mussolini came to power that inflated imperial ambitions began to flourish. By 1934 the two colonies of Tripolitania and Cyrenaica were combined into what we now know as Libya. In 1938 16,000 Italian emigrants were dispatched to the North African territory. Arranged to coincide with the six-teenth anniversary of the Fascists coming to power, ships departed simultaneously from Genoa, Naples and Syracuse. The migrants were mostly of farming stock (although the contingent included shopkeepers and experienced small businessmen) and preference was given to those families having large numbers of children; the *Christian Science Monitor*, reporting from Genoa in November 1938, noted that 'several peasant families had as many as fourteen children'. Libya was to be the granary of the new Italian empire.

Teruzzi by this stage was also fanning substantial ambitions involving the conquered Abyssinia: he was promising Mussolini to plant fourteen million acres of cotton. There was even progress in the largely inhospitable Italian Somaliland: by 1940, there were 179 agricultural enterprises supported by the colony's government with both cotton and bananas being produced in meaningful quantities. The Italians built roads and hospitals.

It was not just imperialistic urges behind the search for colonies. Italy needed to relieve the pressures of its own population, which had grown from twenty-five millions in 1861 to more than forty-three millions by 1935, many of whom were living in poverty. The goal for Libya was to be home to half a million Italians by 1960.

While the British in their colonies had an imperfect record toward the local inhabitants they ruled, efforts were made to help and educate their subjects. The Italians, however, showed little concern for their new colonial populations. While, in 1950, a United Nations report acknowledged that Italian farmers in Libya had executed a remarkable feat of agricultural pioneering and land reclamation, this had been achieved at the expense and suffering of those already living off the land. Herds owned by tribespeople were pushed off their grazing land; in 1926, there had been about 800,000 sheep in Cyrenaica but by 1933 this number had been reduced to around 100,000. Camels and horses owned by the Libyans had been slaugh-tered by the newcomers. It was not surprising then that, when the war began in North Africa and once the British forces managed to push back the Axis forces, the local people attacked the farms of unprotected Italian colonists, notwithstanding British undertakings not to penalise the Italian civilians: a British spokesman said after the capture of Cyrenaica that international convention required any occupying army to observe the status quo and the Italians would not be deprived of their farms. While the spokesman argued that not one of the farms was self-sustaining, and all were subsidised by the Italian government, it was not feasible to return the farmers to Italy, nor could they be allowed to starve. The Senussi tribesman, who had struggled against the colonisation and had been largely suppressed by 1939, rebounded with the disappearance of Italian rule and began taking back their land. By 1944 they had occupied land deserted by Italian farmers and were engaged in growing wheat on it.

However, one point should be made about Italy's African possessions: the British saw Eritrea, Abyssinia and Somaliland as real threats in terms of the Italians using them, or allowing Germany to do so, to attack British shipping and interests in the Indian Ocean. And London believed Italy to be what we now would call a real and present danger to the Royal Navy's supremacy in the Mediterranean and control of the Suez Canal.

## II

By 1942, Italy's economic situation had become dire, partly because of the British blockade in the Mediterranean that prevented the country servicing its traditional export markets outside Europe. Between 1939 and 1945, Italy's GDP fell by a staggering 60.54 per cent.

It was estimated that, by 1942, fifty-five per cent of pre-war export markets had been lost. And, of its European trade, those markets were closed off one by one as Hitler invaded more and more countries. Before the war, Italy had a significant trade with France; after the German occupation, this business largely disappeared notwithstanding the fact that Italy has moved to occupy part of southern France. Trade conducted with Croatia, Montenegro and

Albania (along with Bulgaria) was running as near normal as it could given there was a war on, but these were no substitute for the by now disappeared import-export flows between Italy and the industrialised economies of northern Europe. In fact, the only market (other than Germany) where trade continued as reasonably close to normal was with Switzerland. In the end, most of the rest of Europe was cut off to the Italians. Either various occupied countries were pretty well stripped bare by the Germans for the latter's own survival or, in the case of the unoccupied nations (such as Sweden,) Italy often did not have enough foreign exchange to cover the cost of import needs. And even in markets which they could still reach, the Italians were competing with the Germans; for example, the Nazis had secured from the Vichy government the right to all the phosphate available from French North Africa, phosphate on which Italian farmers had previously depended.

Italy had done little to prepare itself a war economy. The Abyssinian war had provoked sanctions to which fifty-two governments had signed up, thus weakening Italy's mercantile efforts even before it entered the European conflict.

Mussolini, for all his bombast and gestures such as land reclamation and road building, had failed to rejuvenate the Italian economy in the 1920s and 1930s. The Fascist government had in 1933 set up a holding company to acquire stakes in failing companies and prevent them going under; several para-statal organisations became deeply involved in manufacturing and services. But this state control discouraged the sort of enterprise and innovation which a free market stimulates, and workers were demoralised as wages were cut and taxes raised. Nor, unlike the Nazis, did the Fascists manage to make serious inroads into unemployment.

There were some efforts made. Production was stepped up to manufacture lanital, fabric derived from milk casein, to make up for shortages of wool. The Italians called it 'mozzarella' given that garments made with lanital were almost impossible to sew or iron. Output was raised to seven million kilograms a day by 1937. Newspaper accounts of the time noted that wearers of the new fabric

*A wartime Fascist postcard. It reads: 'On the battlefield and on the fields of work to ensure victory for the people's army'.* Author's collection.

complained the garments of which it was made tended to disintegrate when it rained. Lanital was used to make the standard M1934 army greatcoat. These were said to be virtually useless as the material had no warmth-providing qualities. Not that lanital wasn't taken seriously: under the headline, 'One cow equals 40 suits', the *Morning Bulletin* in Rockhampton, Queensland, reported that Australia's Council for Scientific and Industrial Research had termed lanital as 'one of the most outstanding achievements in synthetic fibres' as it dyed more easily and evenly than wool. There were plans to begin manufacturing in France while Princess Cora Caetani, described by the *Baltimore Sun* as a 'noblewoman and fashion authority', in 1939 presented a showing of Italian fashions at the New York World's Fair, telling reporters than Lanital would be a great success with American women.

Italy had no oil production to speak of; and with the country able to produce only about fifteen per cent of its coal needs, this factor prevented Mussolini building a synthetic oil sector as Germany had done (much of the coal was located in southern Sardinia, so there were logistical challenges as well). In 1938, Italy mined just four million tons of coal a year, and imported another nine.

Italy's colony of Libya would not begin producing oil until 1959, long after the Italians had gone, and in 1940 the only reliable source under Rome's influence was the 70,000 tons (just over 510,000 barrels) being produced each year from the oil fields captured the previous year with the Italian invasion of Albania. Efforts were made during the 1930s to convert vehicles to running on either natural gas or wood gas but that made only small inroads into the country's thirst for oil. Until it joined forces with Germany, Italy's main oil suppliers included the United States, Iraq, Central America and Iran; of its other suppliers, only Romania would remain accessible but that country's oil was earmarked by Berlin for Germany's needs. Italy in pre-war years consumed about three million tonnes a year of oil; in 1942 it had to make do with the 1.5 million tonnes shipped from Romania. The situation became so dire that, in early 1941, the Italians were threatening to keep their navy in harbour unless the

Germans could supply at least 250,000 tons of fuel oil. This was not news that Berlin wanted to hear: it was facing its own problems in terms of fuelling its naval fleet. By late 1941 Admiral Erich Raeder, commander in chief of the Kriegsmarine, decided he did not have enough fuel oil available in Norway to send the pocket battleship *Tirpitz* out into the Atlantic. The German naval authorities were resentful, in view of their own shortages, about the order to send some of that fuel oil to Italy. In February 1942 the air raids on the Nazi ships at Brest provoked Hitler to recall them to safer German bases, leading to the audacious, successful and justly famous dash through the English Channel by the battleships *Scharnhorst* and *Gneisenau*, along with the heavy cruiser *Prinz Eugen*. That action alone burnt up 20,000 tons of the Kriegsmarine's stock at that time of 150,000 tons of fuel oil. The Kriegsmarine had begun building underground storage tanks for this oil well before the war. The aim was to have ten million tons stored by 1943—the year German High Command was working toward as to when the country would be ready for war. By 1939, when Hitler's impatience brought on the war four years too early for the military planners, only one million tons of storage capacity was completed (and not filled completely). For all its own urgent needs, the Kriegsmarine between September 1941 and August 1943 delivered 425,000 tons of fuel oil to Italy.

The shortages of oil not only limited the operations of Italy's military forces, it also caused a contraction of the country's manufacturing sector. But lest we be too harsh on the Italians, we should consider the case advanced in 1989 by historian James J. Sadkovich, then assistant professor at the GMI Engineering and Management Institute in Michigan:

> Germany's shortages of raw materials and its failure to prepare for war were more serious than similar Italian failures, given that Germany had provoked the conflict. But while Italy's failures have often been overstated, Germany's have been understated. For example, if Italy had only nineteen of its seventy-three divisions (twenty-three per cent) ready in 1940, the more efficient

Germans had only thirty-five of 103 (thirty-four per cent) ready
in 1939. The Germans also exhausted their stocks of fuel and
ammunition in the attack on Poland, and throughout the war
were dependent on captured and foreign-produced materials to
supply their armed forces—by 1941 even resorting to the use
of Russian guns on Czech chassis. On the other hand, Italy had
attempted to increase its stocks of raw materials and raise industrial production prior to June 1940, but the British blockade
stymied efforts to stockpile.

The minerals situation was not much better. The Fascist regime
did make some efforts: in 1936, it established the Steel Scrap Buying
Cartel but the effort to import more scrap was stymied by the fact
that other countries were competing for what little was available.
Italy was seriously lagging in steel production, its output being
only a tenth of that of Germany's. And Italy was also deficient in
manganese, the mineral vital for steel making and for which—even
today—there is no substitute; it had to import fifty per cent of its
manganese requirements. Lead had to be shipped from Mexico and
Spain; Italy was reliant on the United States and Chile for most of
its copper. The country had no domestic sources of nickel, tungsten
(vital for armour steel and bullets and shell) or molybdenum. And
Mussolini could hardly fall back, as Britain would do, on gold as a
means of paying the import bills; his country in 1937 produced fewer
than 3,000 ounces and could neither match the holdings of the Bank
of England nor, as could London, draw on South African mines to
replenish those. Italy, though did have some metal resources: *The
Economist* calculated it could probably produce enough zinc to meet
its needs, while adequate deposits of bauxite provided the feedstock
for its aluminium production. As *The Economist* summarised in 1937,
'Italy is poorer even than Japan in her domestic supplies of strategic
minerals'. It is surprising, therefore, to find that Italy did pull some
weight in the production of weapons: between 1940 and 1942, the
country turned out just under 2,000 tanks, more than 9,500 military
aircraft (of which nearly 4,000 were fighter aircraft.

Coal was a real problem. Germany was the main source of supply. The problem was that, pre-war, most of the coal bought from Germany came by sea, the rail system being unable to cope. The effect of the British blockade was dramatic when the war began: Germany had delivered 670,000 tons of coal in September 1939, but by November deliveries were down to 175,000 tons with the sea lanes closed off.

Italian railways simply did not have the locomotive and wagons available to suddenly have that seaborne coal traffic transferred to rail. But there was another complication: Italy had for some time bought coal not only from Germany but also from Poland and Czechoslovakia. After the subjugation of those two countries by the Nazis, all the coal produced in those two occupied territories was diverted to Germany's needs. The shortages of coal also had a devastating impact on Italy's steel production: it fell from 2.3 million tonnes in 1938 to 1.9 million tonnes in 1942.

By 1941 not only was Italy militarily dependent upon Germany, but almost thoroughly dependent on its senior partner for trade as well. Between 1938 and 1941, Italy's trade volumes with Germany had quadrupled. There was a strategic as well as economic cost: the growing trade between the two countries placed a severe strain on the trans-Alpine railway system which also had to handle heavy military transport requirements. In 1941 alone, Germany shipped twelve million tons of coal to Italy, along with all the latter's steel import requirements. The great Italian plans for increasing coal production had not materialised: in 1939, the intent had been to treble coal output by 1942. They did not manage that, but they did try. In 1938, the mines in Istria and Sardinia had produced a total 1.37 million tons only. By 1939 the country's mines produced 2.02 million tons of hard coal. At least the Germans kept the price of coal stable; however, by June 1941 the coal shipments started to fall because sufficient rolling stock was unavailable, so that the Italian mines could still not close the demand gap even with higher output. There were reports, too, of Italian Railways complaining that railway wagons sent to Germany to be loaded with coal were being diverted by the Nazis to their own war transport needs.

It was not just industry that was hit by coal shortages. Italians living in the north endued extremely cold winters. No buildings were allowed to be heated before the first day of December and then only for a few hours a day. The Turin newspaper *Gazzetta del Popolo* reported in February 1942 that the Royal Hospital of Charity, which was crammed with wounded Italian soldiers, had been unable to obtain any coal for heating.

Because one of the few resources by which the Italians could pay Germany was food, large quantities, especially vegetables and fruit, were sent north. The Germans agreed in early 1940 to supply specified amounts of coal, benzene, the chemical toluene (also known as methylbenzene, used in petrol, paint and other applications), naphthalene, the solvent acetone and magnesium. In return, the Italians said they would load rail wagons with zinc, bauxite, hemp, mercury, citrus fruit, tobacco, rice and cheese for German destinations.

By 1941 Italians faced severe rationing of bread, pasta, olive oil and cheese. Meat was a rare occurrence in the shops, coffee and tea were not to be seen. What bread and spaghetti that were to be bought was often adulterated to make the basic ingredients stretch further. *Time* magazine reported in November that the only staple food available to Italians in anything like normal quantities was sugar. Before entering the war, Italy imported around 650,000 tonnes a year of grain; in 1941 those imports were down to 85,800 tonnes. Reports from Associated Press in late 1941 told that pure white flour was not be to found in Italy. The 1941 bread ration was 200 grams a day, reduced to 150 grams the following year (although industrial workers, especially those in heavy industry, were allowed greater amounts). Reports reached the British and American press as to price inflation: butter had gone up by thirty-three per cent, beef by more that fifty per cent and eggs had doubled in price. Worse still, wine prices soared by between fifty and 100 per cent. As Edward Townley note in his study of Mussolini's Italy, the calorie allowance under Italian rations was actually lower than set for the Poles under Nazi rule. Food production was further hampered by the conscription for the army of farm labourers and peasants. On top of that Rome buckled to German demands for Italian labour; more than 200,000 men were sent to work in the Reich. The economies of the three of the most important industrial cities—Genoa, Milan and Turin— were disrupted and the devastated by frequent bombing raids by the RAF, these raids not only knocking out factories but sending frightened Italian workers fleeing the cities for the comparative safety of the countryside.

Morale among Italians plummeted as food shortages worsened. Italians, even before the war, had generally poor and low food consumption rates by European standards. The wartime food crisis tipped many poorer people into the a state of malnutrition. The wealthier sections of the community could find ways to buy extra food, but the cost of it meant that many were paying as much as half their incomes to secure adequate food supplies; others exhausted their savings to fend off hunger. What food was produced had to be shared between the civilian populace and the Italian military on various fronts, or shipped to Germany. The time spent searching for food, or being unable to work because you did not have the strength, led to high rates of absenteeism through Italy. Philip Morgan, in his study of Mussolini's fall, quotes a letter from a man in Turin, who tells the recipient that 'every so often Anna Maria calls by to ask for the charity of a piece of bread and most times we don't have any ourselves and cannot give her any'. At one stage the Germans held out the prospect of sending Italy grain from occupied Ukraine, which was seen as an attempt to stiffen Italian resolve as fears of Allied bombing and invasion rose among the population. Nothing came of it.

It was not just shortages of food and lack of any planning to keep the population fed: rationing was poorly organised and corrupt, 'often failing to deliver even the meagre daily caloric allotment of one thousand calories', according to one historian, Michael R. Ebner, charting the level of violence in Mussolini's Italy. It was not so bad if you lived in the country with the ability to grow food for yourself. But the rationing system was further contorted and disrupted as officials took bribes from those wanting to operate on the black market. Small shopkeepers withheld food from supplies that were meant for people using ration cards; instead, they sold it privately for much higher prices. The government in Rome did try and curb the black market: at one stage, the capital was plastered with posters reporting the prison sentence of twenty years for a baker caught selling bread outside official channels. Police met trains arriving in Milan to search country people arriving at the station to see whether they were carrying food to be sold illegally. In the German-occupied north, much food (including eggs and meat) were offered to the military which could afford to pay the high prices demanded.

Between 1929 and 1943 wages in Italy remained static; if they did rise, it was by tiny amounts. Yet prices rose twelve-fold over that same period. As Carol Helstosky writes in her study of food politics, 'across Italy the owners of restaurants and hotels flouted rationing laws; everything was available for the right price'. The average factory worker earned two hundred lira a week; but a litre of wine at a restaurant cost twenty-five lira, a plate of pasta forty lira.

By the end of 1942, Italians were anxious to unload the lira. Cash was ploughed into what was hoped would be safe havens—into hard assets or shares representing commodities or manufacturing materials. The Milan Bourse, as a consequence, saw its index climbing week after week. Between early October and mid-November the volume of share transactions doubled.

### III

If possible, life was to get even tougher after the Italian surrender. Those areas captured by the Allied armies had suffered considerable devastation. First came the British and American bombing and land battles, then followed the German demolitions as the Nazi forces retreated. In Naples, for example, the Germans blew up chemical fac- tories, tanneries, breweries, food canning plants as well as the wharves at the main harbour.

One of the most serious acts of retribution against their former Axis partners was Germany's flooding of the Pontine marshes by destroying the drainage built in Mussolini's time. The marshes,

*The cover of a school pupil's copybook dating from the mid-1930s, headed 'Youth on the March', the name of the fascist youth organisation. These copybooks on their back had rallying messages about the Duce and a quotation from Mussolini himself, this one celebrating a key date in the Fascist "revolution".* Author's collection.

formerly a malarial swamp south of Rome, had been drained at great expense under the Fascist government, and Mussolini regarded the work as his greatest achievement. The impact of the German act was to cause a sharp rise of malaria cases among Italian civilians, an act now regarded as the most serious item case of biological warfare during the Second World War.

Great economic damage was also inflicted by German destruction of electric power generating stations and transmission lines, thus affecting Italian railways and industries which found themselves suddenly without electricity.

# 8.

# The slow economic strangulation of Germany

ON 21 APRIL 1944 Turkey's foreign minister announced his country was ceasing exports to Germany of chrome, a metal vital to the Reich's steel and weapons industries. This left the Nazis with only about half the chrome they needed, and Greece and Yugoslavia as their only sources of the metal. Chrome was needed to toughen steel; the only substitute was manganese, of which there was also serious supply problem by this stage of the war. How times had changed. In the early years of the war, the Nazis had ready access to Turkish chrome: in late September 1942, for example, the Turks signed a deal to ship 45,000 tons to the Krupp Armament Company.

German industry by this later stage of the war could still count on iron ore supplies from France and Sweden but, apart from chrome and manganese, the Reich was heavily dependent on supplies from beyond its 1939 borders for its needs of tungsten, nickel, cobalt, molybdenum and vanadium, all metals that played a part in turning iron ore—a soft metal—into strong steel. Tungsten, for example, was vital for the strength needed for machine tool manufacture; the metal was also used in light bulb filaments and radio and radar tubes.

Turkey had by this time been subjected for several years to pressure from both London and Berlin on the question of chrome, and by 1944 the Turks could see the writing on the wall for the Nazi regime.

At the onset of war in 1939, Britain and France had contracted to take all the chrome Turkey could produce. Then in October 1941, Germany and Turkey signed a trade agreement which provided the two countries would exchange about 200 million marks worth of goods over the following eighteen months. The German communiqué mentioned exports of cotton, tobacco, olive oil and minerals, especially the breaking of the British monopoly on chrome exports from Turkey—the country being the then largest producer of the metal. In September 1943, for example, it was reported Germany had delivered fifteen locomotives and several hundred railway freight wagons to Turkey in payment for chrome.

But British pressure on Turkey continued; as did the military pressure from Soviet forces. In January 1944 the German freighter *Thisbe* was hit by two torpedoes after leaving Istanbul and in a position sixteen kilometres into the Bosphorous. It went down with 2,000 tons of chrome in its holds. It was surmised at the time that a Soviet submarine was responsible.

It was not just Turkey that had been subject to Allied lobbying to keep neutrals in line: the Spanish Government had agreed in 1944, after coming under pressure from Washington to reduce its tungsten exports to Germany from 100 tons a month to twenty tons. From the moment it entered the war, the United States bought as much Spanish tungsten as it could on the open market, this buying forcing Germany not only to pay more with scarce foreign funds but to use less, a critical blow given tungsten's importance as a hardening agent in machine tools, ammunition and weaponry.

Oil was a very useful weapon, most notably in the case of Spain: the Allies used their control of oil supplies to the Spanish by means of the naval blockade of Europe to pressure Madrid not to join the Axis powers. The British were able to put what was called a 'squeeze' on supplies to bring the Spanish to heel; they also made sure that aviation gasoline supplies were strictly limited. The most tangible results of Allied policy, apart from the primary goal of keeping Spain neutral—and limiting the material aid sent to Germany—included forcing the Franco regime to withdraw its Blue Division from the eastern front; this all-volunteer unit attracted many veterans of the civil war and members of the Fascist organisation, the Falange. They were trained in Germany and fought in several battles, including the siege of Leningrad. The Spanish, some of whom had never seen snow before, were sent into battle in the middle of the northern winter and suffered considerably. The legion was withdrawn in October 1944 due to a combination of pressure from London and Washington and growing disquiet within Spain.

But, initially, the Allies were not quite so successful in stemming Spanish exports to Germany of key commodities; after all, it was not just sympathy with the Nazi cause involved but Spain (like Italy) had lost many of its export markets either due to the naval blockade or the fact that Germany occupied many of their former customers.

Germany was, in 1938, the world's largest consumer of tungsten, accounting for a quarter of global use. It was, after oil, the com-modity Germany wanted more than anything. It has the highest melting point of any metal and is very hard—hence its importance to armour-piercing shells, for example, and the machine tools that produced Germany's aircraft, tanks and munitions. Berlin's problem was that the bulk of supplies had until then come from mines that by 1939 were cut off. In 1938, Germany's main source of tungsten had been China, the Nazis importing 8,037 tons that year; another 1,229 tons came from India. A long way behind as sources were Portugal (304 tons) and Spain (150 tons). Britain first put pressure on Portugal to ensure they themselves and the United States would get shipments of the vital metal; until 1941, most of Portugal's exports of tungsten had been going to Germany. But it was Spain that was to benefit mostly from the competing bids for its tungsten. In 1940, Spain earned the equivalent of just £73,000 from its tungsten exports; by 1944 the foreign exchange income from the metal had risen to £16.26 millions, thanks largely to the bidding war between

Britain and Germany which resulted in astronomical increases in the metal's price. By 1943, Spanish tungsten was costing the Nazis 141 million Reichsmarks more than it had in 1938, a serious impost on German finances.

Britain had a very powerful economic weapon when it came to dealing with Spain, one that Germany could not match: the problem was that Spain had no domestic oil resource, and its only refinery was on the Canary Islands which could be blockaded by the Royal Navy. In 1938 the Germans had attempted to seal a deal with the Franco government in Madrid. They proposed making German tankers available to bring Mexican oil to the Canary Island refinery, in return for which Spain would allow U-boats to refuel at the islands.

For Britain, the Spanish question was also about Gibraltar. Had Germany and Italy gained control, with the help of Spain, to 'the rock', it would have cut off the Mediterranean to the Royal Navy. As it was, Winston Churchill had voiced fears that, if Spain joined the Axis, there would be howitzer shells raining down upon the British naval base from across the Spanish border.

The problem for Franco was that Germany could not meet its obligations of supply Madrid with key supplies; requests went to Berlin for oil, wheat, coal, cotton, rubber and other items. Germany, of course, was struggling to meet its own needs. So Madrid was forced to deal with the British, who could help. (As for the wheat, Canada and the United States would step up to that plate.) The price? Neutrality. The Spain counter was taken off the Axis side of the table not by military strength, but economic muscle.

Tungsten had kept moving to Germany from Spain in spite of the Americans offering good prices for Spanish output. Then, in 1943, Washington began to tighten the vice: Washington demanded further Spanish cooperation in return for oil, then in January 1944 threatened Madrid with a total oil embargo unless the Spanish handed over Italian ships stranded in their ports as well as reducing tungsten exports to Germany. Similar pressure was brought to bear on Portugal which had been exporting tungsten to Germany in return for steel and fertiliser. Lisbon was also brought to heel by threats of cutting off oil supplies, but it was not until Allied troops landed in Europe that the Portuguese became fully cooperative on the tungsten question.

## II

One powerful economic power on the Allied side tends to be overlooked: the Union of Soviet Socialist Republics. Much attention—including, of course, in this book—has been given to the extraordinary resources the Americans were able to bring to bear on the war effort, expanding and harnessing their industrial complex, money and raw materials. Yet once the Soviet planners recovered from their shock at the German invasion they were able to pull all the levers of a command economy, stepping up Russian production and that would far outstrip Germany's industrial efforts.

By 1944, at which time German factories were being bombed regularly and raw materials were often in short supply, and the Luftwaffe was fighting a losing battle due to fuel shortages and loss of aircraft and aircrew, Soviet factories were just hitting their straps. In 1944 four times as many aircraft were being produced by the Russians as had been delivered in 1940. In the case of armoured fighting vehicles, production had increased ten-fold. With machine pistols the factor was twenty-four fold. Moreover, the quality of Soviet weaponry improved as the war progressed.

But there the Soviet story departs from the American one. In the former's case, as Mark Harrison illustrates, much of this additional weaponry was wasted.

> Excessive losses arose in the early period of the war because of the incorrect use of tanks, deficiencies of leadership, and the lack of spare parts. The wasteful deployment of tanks units continued through 1942; even after 1943, when fully motorised and independently operating tank formations were created, they continued to be used inappropriately, for example, for assaults on large cities right through 1945.

Yet even while Luftwaffe pilots shot down large numbers of Soviet aircraft, it did not make all that much difference in terms of war output. German bombers—those that were left—were not able to attack the factories producing replacement fighters and bombers because those plants were located far from the front. (The eastern front did offer one opportunity for the Germans: the average skill of Soviet pilots was so far below that which obtained in the Royal Air Force or the United States Air Force that the Luftwaffe used the Russian front to get its pilots experienced, and only then were they transferred to take on the far more daunting British or American flyers.)

In addition, it tends to be overlooked that Britain was no shrinking violet in rebuilding its industrial capacity. The 'plucky little Britain' image is correct—up to a point, as Lord Copper was told. As recorded by the Ministry of Information in 1945, at the beginning of the war British factories were turning out 730 aircraft a month; by 1943, this rate of delivery had trebled, but was sixfold as measured by structural weight (a Lancaster bomber obviously rating more than a trainer); in the first six months of 1944, 2,889 heavy bombers were turned out in British plants compared with only forty-one in the whole of 1940. Germany, meanwhile, was struggling in 1944 to replace combat aircraft being shot out of the sky. Indeed, over the duration of the war Britain produced 102,609 military aircraft. Given that a single aircraft could have up to 70,000 pieces and shapes of fabricated materials, this was an astonishing achievement. The average RAF bomb load increased from 1.2 tons in 1939 to four tons in 1943. Bombs in use in 1940 weighed 500 pounds; in March 1943 4,000lb bombs were being produced in Britain. A year later factories were turning out 12,000lb ones.

Not too much went smoothly for German efforts to harness the resources of its conquered territories. Berlin wanted not only that the state-owned French aircraft factories produce aircraft to supplement the Luftwaffe replacement programme, but that Germany take ownership of those factories. As it transpired, Germany had to forgo the ownership move in order to get Vichy agreement to turn out 3,000 aircraft for the Luftwaffe along with 13,500 aero engines. As Adam Tooze points out his his study of the German war economy, France had the bauxite resources and the smelting capacity to make the aluminium necessary for the airframes, but needed Germany to deliver 120,000 tons of coal a month in order to generate the electricity needed for the smelters. Germany could guarantee only 4,000 tons a month.

By way of contrast, as Britain entered 1942 it had already received 5,012 aircraft from the United States (it would get many more in 1942) while France and Holland (Fokker) delivered just seventy-eight military aircraft between them to the Luftwaffe, as a study by Ference Vaida details. In 1930, France had possessed Europe's largest aircraft industry and, although overtaken by Germany, was by the declaration of war still a significant producer. Under the agreement between Vichy and Berlin, two-thirds of aircraft output was to be tagged for Germany (the remainder went to the French Air Force). While output lifted in 1942 (to 300 aircraft a month) the French plants were subjected to frequent bombing raids which limited production. Plans were drawn up to build underground plants to turn out the Focke-Wulf Fw-190 single-seater fighter but nothing of the scheme had been implemented by the time of the D-Day landings. The main production achievement of Vichy factories seems to have been the completion of 602 of the Ju-52 tri-motor transport aircraft. While captured factories in Czechoslovakia turned out many aircraft for the Germans, United States Army Air Force bombing thwarted any attempts to have Hungarian and Italian plants harnessed to the Luftwaffe requirements.

The war saw British shipyards launch 722 major naval vessels; again, this has to be seen in the context of a war policy that new ships were given lower priority than the production of aircraft and army equipment (in the case of the latter, an acute need after the losses at Dunkirk). But then London also had to worry about the merchant shipping being lost to the U-boats; gross tonnage of cargo vessels launched between September 1939 until the end of 1943 was 4,717,000 tons.

Iron ore mined in Britain rose fifty per cent during the war, and steel production was obviously up. Yet as early as 1940 reports were coming out of Germany quoting an an interview in the *Voelkischer Beobachter* newspaper with Göring in which the Nazi leader requested Germans deliver to the army 'every article made of copper, bronze, brass, tin, lead and nickel that is not vitally needed by the citizen'.

## III

The Germans lost 3.5 million soldiers killed on the Eastern Front; another three million were captured by the Russians. While some ten million Soviet troops also died, Moscow still had many divisions available to it. Berlin did not. The numbers of dead still challenge the mind's comprehension. No wonder the German invasion, and defeat, in Russia and Ukraine have been the subject of so many books; and, in Russia's case, so many motion pictures—memorable films from *Ballad of a Soldier* to *The Cranes are Flying* and *Come and See*. But there is a part of the Eastern Front story that has had little recognition: the implications of what Germany lost, and what the Soviets regained, as Hitler's armies were pushed back.

This was the second time that Germany forewent the immense economic benefits of being on friendly terms with the Russians. Under the now infamous non-aggression pact signed in 1939 between the two foreign ministers, Joachim von Ribbentrop and Vyacheslav Molotov, the economic benefits flowed both ways, although Germany got by far the better of the deal. Up to the moment German troops began attacking Soviet forces in 1941, Berlin shipped about 6,500 machine tools to the Soviet industry, which was seen as vital to meeting targets in the Third Five-Year Plan, the half-completed heavy cruiser *Lützow* (renamed *Petropavlovsk*), combat aircraft including two Dornier Do-215 light bombers, five Messerschmitt Me-109s and Me-110s, two Junker Ju-88 dive bombers, five Heinkel He-100 fighters, along with three each of the Bücker Be-131 and Bücker Be-133 biplane trainer aircraft.

But the traffic the other way was far greater. By mid-1940, the U.S.S.R. was Germany's most important supplier of raw materials. The Russians sent Germany 1.7 million tons of grain, one million tons of fuel, along with cotton, manganese, chrome, phosphate, asbestos and wood. As noted elsewhere, the Russians also allowed the use of the Trans-Siberian Railway for the carriage of German imports from suppliers on the Pacific Rim. To place this in some sort of historical context, this boost to trade between the two was actually a reversion to the *status quo ante*; that is, in the 1920s Germany was the Soviet Union's most important trading partner; it was only the rabid anti-communist policies of the Nazi government which in 1933 had led to disruption of much of that trade.

Once the Red Army was able to stop the Germans in their tracks at Stalingrad, Kursk and elsewhere, the Soviets were able to replenish their battalions and labour force with men now available from the newly liberated areas, there was a rapid rejuvenation of industries not only in the reconquered zones but others that remained behind Soviet lines and had been disrupted by the fighting or threat of it. By May 1944 petroleum production in areas liberated was ahead of what Moscow had deemed in planned quotas, and the coal and iron and steel industries of the Donets Basin of eastern Ukraine were once again operating, although on a much reduced scale from their pre-war state.

In the twelve months to May 1944, according to the League of Nations, the Soviet Union's pig-iron production increased by twenty-seven per cent, steel production by 26.5 per cent, rolling-mill production by twenty-three per cent, and coke production by thirty-nine per cent. 'These increases reflected not only the restoration of productive capacity in the liberated areas, but also the continued expansion of production in the new industrial areas in the Urals and further east,' the league reported in 1945.

By contrast, Germany's military manpower losses forced her to intensify the mobilisation of labour in 1943 and 1944 for the purpose, not of increasing her war effort, but of preventing a precipitate decline in its industrial output. Germany may have talked 'total war' in terms of the economy, but they took their time walking the walk, so to speak. Following the defeat at Stalingrad in 1943, what was called 'total mobilisation' was declared; yet this still left room for a further such declarations with more stringent measures after the Normandy landings. The post-Stalingrad edict required all men aged sixteen to sixty-five, and all women between the ages of seventeen and forty-five not already in the military, to register for compulsory labour; its most apparent effect was to force the closure of large numbers of retail stores. More foreign workers were brought to Germany, and in 1944 there were seven million of them engaged in German industrial production (about twenty per cent of the country's factory workforce) but this relocation of foreign workers was not easy at this stage of the war (for logistical reasons) which led to a changed policy for keeping workers in their own countries but under military control. That worked only so long as Germany occupied those countries; as Hitler's armies were pushed further and further back toward their own borders, with that retreat went the foreign workers who now found themselves back in Allied-held territory. This meant Germany having to put further pressure on its own nationals. Joseph Goebbels, as well as being Propaganda Minister, was given the title of Reich Trustee for Total Mobilisation for War. All entertainment venues, apart from cinemas, were closed; now women between forty-six and fifty were drafted to work; many newspapers were closed and only political and technical publications were allowed to operate. But the well of non-essential activities that could be closed in order to free up labour was getting very low.

Apart from enemy armies on two fronts pressing toward Germany, there was the added problem of the air raids. Industries had to be dispersed, with all the dislocation of production, even if only temporary, which that involved, as well as putting extra strain on an already hard-pressed railway network. Workers who should have been on the factory floor had to spend valuable time repairing bomb damage, not to mention the dismantling and rebuilding required. On top of that, it was usually not possible to achieve the same production volumes at the new sites, especially if the factory was located in a rural area rather than an industrial city. Disused mines and railway tunnels were harnessed to house vital war plant.

But there were some operations that could not be moved, synthetic oil plants among them.

Then, with the liberation of France and the Balkans, Germany lost food supplies that had helped keep its own citizens and troops fed. But the enormous burden on the railway system meant even food produced within Germany often could not be moved too great a distance, so increasingly most of it was consumed closed to where it was produced rather than where it was most needed. Much has been made of the extraordinary achievements of Albert Speer as Minister for Armaments and Munitions; that much is true, and his reorganisation of industry and allowing individual managers more control over their own operations certainly sustained the German war effort and allowed the fighting to continue. But there was a time limit even to that simply because so much else in the Reich economy was falling apart. Moreover, British and American production was increasing at an even faster rate.

Really, 1944's problems were only an extension of failings that had been obvious even when Germany's temporary empire was at its zenith. The inability of the Germans to improve transport within its conquests defeated efforts to meld those economies with its own, in spite of tariff walls being abolished with Belgium, Estonia, Denmark, France, Greece, Latvia, Lithuania, Hungary, Italy, Norway, the General Government (Generalgouvernement in German, the part of occupied Poland governed as a German colony), the Ukraine and Yugoslavia (except areas occupied by Bulgaria). There was never any hope of establishing a system of trade that made the extended Reich a coherent economic entity. Trading tended to be done within each occupied territory rather than between them, partly because there was minimal ability to provide transport for any non-military goods.

The trade from French North Africa had been lost by late 1942. That region had previously exported just under 400,000 tonnes of wheat a year. Critically, this was also the source of Europe's phosphate for fertilisers. As the League of Nations reported, North Africa had been a source for Germany of wine, olive oil, grapes, dates and citrus fruit. Another source of the last mentioned also was cut off once the Allies occupied southern Italy, Sicily and Sardinia, as were supplies of pyrites, molybdenum, sulphur, zinc and antimony. As we have seen, Portugal and Spain exports of tungsten had been nobbled, and Turkey's chrome shipments likewise. The Swedes decided they could now defy Berlin and refuse further supplies of ball bearings.

The Portuguese and Spanish decisions were soon made moot. The liberation of France cut off trade links between the Iberian peninsula and Germany. The loss of trade with south-eastern Europe deprived Germany of mineral oil, cereals, oil seeds, tobacco and many other products. Between November 1942 and December 1944, the trade volumes for Germany and Austria with occupied and neutral Europe declined by about two-thirds.

## IV

But, back to the beginning. In the mid-1930s, German High Command laid out a plan for what was known as 'rearmament in depth', building both a highly mechanised military machine and an efficient production system to re-supply its forces. Yet, at the outbreak of war, the Wehrmacht counted as part of its fighting machine 590,000 horses. By the time of the invasion of the Soviet Union, the armies poised to put Operation Barbarossa into action had 625,000 horses. These animals actually impeded the advances into Russian territory, which were dramatic in the first weeks of the action, because horses had to be stopped and watered on a regular basis; and vital transport space was taken up by the need to ship large quantities of fodder to the front.

The High Command saw Germany being ready for war not before 1943. Hitler could not wait. But, again, economically rational thinking played no part: as a British government report compiled after the war by the Joint Intelligence Sub-Committee argued, the Nazis refused to allow any fall in domestic living standards even with their avowed 'guns before butter' proclamation. Thus the German economy was placed under an unnatural strain from the offset, struggling to both expand at a fast clip the military supply chain and also not snatch away the advances made in terms of prosperity since the Great Depression ended. The Americans had the resources to pull off this rather tricky feat, the Nazis did not.

At first, the strains did not show. Norway, the Low Countries and France were invaded and conquered without any apparent shortfalls in materiel. Later in the war, Albert Speer as armaments minister achieved extraordinary results: by the middle of 1944 he had factories producing more than three times the quantity of armaments than they had in 1942. That achievement is well known and documented but it begs the question: why did not Germany achieve that level of output in 1942, a feat which could affected the course of the war? That very omission lay in the failures before the war. The tide turned in Russia by early 1942. Had the armaments industry been then in at a level of output as obtained in mid-1944, Germany might just have been able to supply replacements for the vast amount of equipment that was lost in the reverses on the Russian front and at least slow the Wehrmacht's retreat. But it simply did not have the capacity.

## V

On 19 January 1942, Russian troops re-entered the town of Mozhaysk, 110 kilometres west of Moscow on the road to Smolensk and Poland. The Soviet troops found hundreds of Germans who had been frozen to death; for once, the town had not been destroyed by the retreating Germans, the Russians having arrived before the usual setting of the fires had begun. It was just part of the great battle front as the Soviet forces had begun to turn back the Wehrmacht. In the writing of the history of the war, and of the eastern front in particular, there were so many battles of epic proportions—Stalingrad, obviously, the siege of Leningrad, the battle at Kursk where the Germans lost 252 tanks in one action—that the re-occupation of Mozhaysk is just another minor detail. But there were thousands of minor incidents such as this, and each one dealt an economic as well as military blow to the Nazis.

From 1941, the Royal Institute of International Affairs used its journal to publish regular updates on the war under the heading 'Outline of Military Operations'. In the issue of 7 February 1942 there were listings of cities of Germany that had been bombed over the past fortnight by the Royal Air Force, details of the Russian claims of German losses over December and January (2,766 tanks, 4,801 field guns, 3,071 mortars, and 33,640 cars), news that the air attacks on Malta were continuing, confirmation of the loss of battleship *HMS Barham* with 850 lives, news of a fresh advanced by Rommel in North Africa along with General Douglas MacArthur's troops defending Bataan in the Philippines, the defence of Singapore and the shooting down of fifteen Japanese aircraft over Rangoon.

These bulletins would continue over the next three and a half years to chronicle the great ebbs and flows of the Second World War.

Those bulletins embody the type of information about the Second World War which, to a greater or lesser extent, consists of the average knowledge of the conflict. But the economics of the war remain largely obscure; and they were not subject to regular updates.

## VI

As Adam Tooze points out in his magisterial account of Germany's war economy, it was not the number of women in the German workforce that counted, it was what they did. In 1939 a third of all German women were economically active, compared with a quarter in Britain. Yet, of Germany's fourteen million women workers in 1939, only 2.9 per cent of them worked in industry; six million were engaged in peasant agriculture. That same year, only 100,000 of the six million working women in Britain worked on farms.

With men, the problem was that the enormous needs of the Wehrmacht for more recruits caused production problems in the industries from which they were removed. Between May 1939 (when the army already had more than 1.1 million men in uniform) and

May 1940, the Wehrmacht called up around 750,000 farm workers, 1.3 million industrial workers, 930,000 craftsmen, 220,000 transport workers, 600,000 retail workers and 600,000 clerks. While many of those recruited could continue to employ their skills for the overall German war effort—and that was particularly the case with skilled industrial workers such as fitters and electricians—those farmers and farm hands pulled off the land could no longer help German produce enough food to feed itself.

But there is one extraordinary fact that goes to the sometimes haphazard processes that passed for economic planning in Nazi Germany. Until 1943, there was no central registration of labour for national service. Labour conscription was introduced in 1939 but was not used widely: by September 1942 fewer than 700,000 German men had been directed to report to new places of employment. When the registration system was finally introduced at the beginning of 1943, it covered men aged between sixteen and sixty-five years and women between seventeen and forty-five years of age.

## VII

Germany, of course, had one advantage over the Allies: it did not need to cajole, bargain or pay full price for much of what it wanted. It just took, and took and took, from the occupied territories. In 1942 the League of Nations, in its economic survey of that year, reckoned that what it called the 'financial exactions' from France, Belgium, Holland, Norway and Denmark were of the order of ten per cent of the German national income. Out of a total German government budget of eighty-eight billion marks, some seventeen billion represented unwilling foreign contributions.

The exploitation of the occupied territories worked only as well as their individual economies which, in almost all cases, was not well. The Netherlands, for example, was a highly developed economy by the time the Germans invaded the country. It was the home for several multinationals (Shell, Philips and Unilever being the largest) but there was also a substantial middle-sized manufacturing sector.

Yet many of those factories closed in the winter of 1941-42 because of coal shortages and did not re-open until the war was over. In 1942 Dutch industrial production fell 17.2 per cent. Berlin, in the form of the Minister of Armaments and War Production, Albert Speer, ordered the closure of all small companies in the Netherlands, and all companies producing what were considered non-essentials. This was aimed both at saving raw materials and also freeing up labour for war work; the workers of the companies closed were sent to Germany. Eighty-seven per cent of clothing and shoe manufacturers were shuttered, eighty-nine per cent of Dutch soap factories, seventy-nine per cent of paint manufacturers and even a third of the shipyards.

German industry needed to become more efficient. The armaments drive of the 1930s, with all the government control and direction involved, had stifled the competitive urges and freedom in companies. Bringing all the labourers from occupied territories to Germany (they tended to be young and the more skilled workers) not only affected those territories' economies, but made them close to worthless to the German war effort. In France, apart from the million men taken to Germany, another million went into hiding, removing that number from the workforce. More than 300,000 Dutch also went into hiding.

Then there were the unintended (or, least, seemingly ignored by Berlin) consequences. The Netherlands Indies before June 1940 had been an important source of raw materials for Germany, so much so that when Holland was defeated there was still a considerable number of German freighters loading cargoes in the Dutch colony's ports; as is outlined elsewhere, the non-aggression pact between Hitler and Stalin meant Vladivostok was an important receiving centre for goods bought by Germany from various parts of the Pacific Rim, goods which were then transhipped to Germany via the Trans-Siberian Railway.

But the invasion of the Netherlands by the German forces on 10 May 1940 'raised a wave of horror, indignation and fury' in the Dutch colony, according to a despatch published five days later in *The New York Times*. The authorities in Batavia (now Jakarta) ordered the immediate arrest of all German nationals and even Dutch citizens suspected of Nazi sympathies. The first arrests were made within two hours of the invasion news reaching Batavia. All German firms were closed, their premises placed under seal. About nine hundred Germans were placed in prison camps and, more importantly so far as the German war effort was concerned, nineteen German merchant ships were seized.

In late March 1940 it had been reported that a number of German vessels had been loading at ports around the Dutch-held archipelago. These ships had been stranded in the Indies since the outbreak of war the previous September. Many of the ships's crew had sold off items to meet their financial needs, but by March it seemed that they intended to make a dash for it, presumably toward Vladivostok. At Batavia the *Nordmark*, the *Rendsburg* and the *Vogtland* had been taking on mixed cargoes including rubber and pepper; in Surabaya, kapok and copra was being lowered into the holds of the *Cassel*, the *Essen* and *Naumburg*; meanwhile, at Padang the *Soneck* was laden with rubber and other goods. *The Times* reported that Dutch sources were saying the ships were planning to sail for Vladivostok and it was 'not doubted that some of these supplies were intended for Germany'; it was assumed that the goods would then be moved across the Trans-Siberian Railway. In Singapore, the *Straits Times* was reporting that the German ships were noted to have taken larger than usual quantities of coal and the vessels were being repainted grey with their names removed. The report went on to state that all the German purchases had been paid in cash. But the ships were seized before they could set sail. One more source of supply for Germany was thus cut off.

## VIII

The term 'Battle of the Atlantic' has left a haunting cultural memory of all those British and Allied ships that were sunk, and all those thousands of men who lost their lives, often in the most appalling circumstances. But two points deserve to be made. One, while the horrific losses to Allied—particularly British—merchant shipping did dent the anti-Nazi war effort, the Allies never actually ran short of any critical supplies to the extent that those losses caused any serious fear on the Allied side. It was quite the opposite situation for the Axis powers as German and Italian merchant shipping suffered huge losses.

But, first, the British experience, for which we should just look at one example—Britain's Ben Line.

The Ben Line vessel, *Benwyvis* (5,920 gross tons) was torpedoed on 21 March by U-105 north of the Cape Verde Islands. She was part of Convoy SL68 bound for Britain out of South Africa, this vessel having originated at Rangoon. Thirty-four of her crew were lost. This is not an unusual story in the Atlantic during the war. The merchant marine in the Atlantic was one of the most dangerous places to be: a slow-moving freighter with inadequate defences, was a prime target for the U-boat packs.

But while the anguishing images of men flailing and drowning in what novelist Nicholas Monsarrat rightly described as 'the cruel sea' were seared on post-war memories, the sinking of such ships as the *Benwyvis* had serious implications for both the shipping lines that owned them and the economic survival of Britain; while Britain, as we have noted, still received enough cargoes to carry on, there was a real fear (and possibility) that would not be the case. For the Ben Line, this was a dreadful period of the war. Its 5,800-ton freighter *Benarty* was captured on 10 September 1940 by the German raider *Atlantis* in the Indian Ocean and then sunk using time-bombs; two days later, the 5,872-ton *Benavon* was sunk by the merchant raider *Pinguin*, also the Indian Ocean. Not a month passed before the *Benlawers* was lost, the vessel having become a straggler from Convoy OB-221 travelling from Swansea to Port Said via Cape Town carrying army stores including trucks, and being hit by a torpedo from U-123.

On 22 March 1941, the day before *Benwyvis* met its fate, it was the turn of *Benvorlich*, 5,193 tons, which was carrying explosives from Britain to the Far East—she was attacked by a Focke-Wolf Condor and a single bomb caused the ship to explode and twenty of the crew to be lost. Other Ben Line losses were the *Benmohr*, sunk by U-505 on

5 March 1942 while en route from Bombay to Oban; the *Benlomond* torpedoed by U-172 on 23 November 1942; the *Benmachdui* which hit a mine in 1941 while carrying supplies destined for the Royal Navy in the Far East; the *Benvenue*—another U-105 victim—was hit by a torpedo 675 km south-west of Freetown while travelling from London to Karachi, going down on 15 May 1941 with its general cargo and six aircraft. The *Bennevis* was captured by the Japanese when they seized Hong Kong and was renamed *Goyoku Maru* (and subsequently sunk by a United States submarine in 1944). The 7,153-ton *Benalbanach* was hit by an aerial torpedo off Algiers on 7 January 1941 while carrying troops to North Africa, going down with a loss of 410 people (including 353 men of a motor transport unit destined for the front). The *Bencruachan* hit a mine off Alexandria and sank on 5 July 1941.

Setting aside the horrific loss of life in each of these cases, each sinking of a merchantman deprived the Allied war effort of vitally needed supplies. *Benwyvis* went down with 3,500 tonnes of rice, five hundred tons of lead, 1,100 tonnes of timber and fifty tons of wolfram; *Benmohr* was carrying silver bullion, pig iron and rubber. With other shipping lines, when the *SS Belgravian* was torpedoed off Ireland on 5 August 1941 by U-372, the Allied war effort was deprived of 3,946 tonnes of kernels, groundnuts and tin ore from Nigeria. As for other shipping lines, *SS Tasmania* was carrying 8,500 tons of food, tea and jute along with two thousand tons of pig iron and four hundred tons of iron ore from Calcutta to Glasgow when hit by a torpedo from U-103. The *SS River Lugar* took 9,750 tonnes of iron ore to the bottom after being hit by a torpedo from U-69; then U-106 deprived Britain of 3,231 tons of cocoa which was aboard the *Andalusian*. The *Port Auckland* was on the last leg of her voyage from Brisbane when, in the mid-Atlantic on 17 March 1943, she was struck by a torpedo from U-305 and went down with 7,000 tons of frozen produce along with 1,000 tons of general cargo and mail. Another 7,840 tons of frozen meat and 5,000 tons of general cargo from Buenos Aires was lost on 6 September 1942 when, after a call at Freetown, the *Tuscan Star* was spotted by U-109.

But, it should be recorded, there was another side to this story, and one that has tended to be overshadowed by the 'Battle of the Atlantic' that has been so widely covered in books and films.

Germany and Italy also suffered grievous merchant shipping losses. By the end of 1943, the two Axis powers had seen some 10 million gross tons of their merchant fleets end up on the sea floor. In fact, the losses totalled more than the tonnage these two powers had possessed before the war, but Germany particularly had acquired many ships belonging to countries it invaded. Even then, German shipyards—despite the pressure to re-equip the German Navy—had to start building new freighters in 1943 to have enough capacity to maintain shipping links with occupied Norway once Sweden closed off its rail system to the Nazis.

On 5 August 1943, the Swedish Government announced it had cancelled the arrangements with the German government to allow Nazi troops to cross its territory. This brought to a close a less than proud moment in Sweden's record of neutrality. When the Germans invaded Norway in April 1940, the Swedes initially had imposed a ban on assistance to both sides in the conflict—no war materiel could be transported across its territory for either the Germans or the Norwegians. Once the invasion was completed and Norway subdued, however, the government in Stockholm announced (on 6 July) it would allow German soldiers 'on leave' to travel across its territory, a move motivated by Sweden's fear that Germany would turn on it if such a concession were not made. Moreover, Sweden was heavily reliant on Germany for its coal supplies. The agreement meant troops returning to Germany and their replacements could avoid transport via the North Sea coast which carried with it the risk of attack by British aircraft and warships. Daily trains travelled between Norway and the Swedish port of Trelleborg, the closest port in southern Sweden to the German coast.

The agreement stipulated that the Nazi troops could not carry weapons other than side-arms while in Sweden, but this was sub-verted by storing larger weapons in special vans attached to the trains. Moreover, travel by German military personnel across Sweden to and from Finland was specifically prohibited, but by 1943 it had become public information that the Germans were using two passenger carriages attached to trains that travelled between Storlien, close to the Norwegian border, to Haparanda on the Finnish frontier at the northern end of the Gulf of Bathnia. This particularly riled Swedish public opinion because it was seen as allowing Germany the right to move troops between two fighting fronts—Finland being at war with both Britain and the Soviet Union—rather than maintain the fiction that it was merely for soldiers going on or returning from leave. It was reported in British newspapers that heavy artillery and fuel for the Luftwaffe in Norway were also sent over the Swedish railway system. The Swedish prime minister Per Albin Hansson could go only so far to describe the arrangement as a 'burden'.

Earlier in the year, newspapers and trades unions in Sweden began a campaign to end the transport of the Germans. They argued that Sweden was in no position to resist German demands in 1940 but, by 1943, the Nazis could not realistically mount an attack on Sweden.

*Time* magazine reported in May 1943 that the movement of Nazi troops through their country was 'the sorest spot in Sweden'. Thrice weekly, it said, Nazi troops were carried by special trains between Storlien and Haparanda. These 'Reichswehr specials' were guarded by Swedish troops. 'The sight of well-fed Germans hanging out of train windows, yahooing at Swedish girls, and carrying packages of food, butter and herrings out of starving Oslo is almost too much to stomach,' *Time* said.

Possibly one of the best known facts about Sweden and the war was that it was a critical source for Germany of iron ore. But, in fact, the country suffered a loss of about seventy per cent of its export markets once Denmark and Norway fell into German hands. The Swedes, after long negotiations with both Berlin and London, managed to get permission for some ships to pass through the British blockade and German-controlled waters with vital supplies of petrol, oil-cake, raw material for its textile industries and other much needed imports. While the Swedes were selling iron ore to the Nazis, they relied in return on Germany for coal, heavy chemicals and industrial products. Initially this worked reasonably well, but only up until the invasion of the Soviet Union coupled with the increasing air raids on German industry by the RAF. This meant Germany had less coal and fewer industrial products it could spare for Sweden. But petrol was in such short supply that those living in Swedish towns soon became familiar with the sight of lorries, buses and even taxis powered by wood or charcoal burners. And householders were allowed coke only in the harshest winter months; for the rest of the year, they were forced to use only wood for heating. However, and unlike in most other parts of Europe, the Swedes still ate comparatively well; a number of items were available only with ration cards, but fish, potatoes, milk and vegetables were still freely available.

## IX

That was unoccupied Europe. But Germany did not have it all its own way in parts of continental Europe it had subjugated even during the first years of the war. Once the Nazis occupied France and the Low Countries, and Italy joined the fight, Britain's communications with neutral Switzerland would seem to have been disrupted. Not entirely, as one study by University of Nottingham historian Neville Wylie demonstrates. British smuggling out of Switzerland was successful to the tune of £1.8 million worth Swiss contraband by October 1944. But, as we have seen, Britain started making preparations well before the war began and in June 1938 placed orders for fuse mechanisms with a company in Geneva; then came orders for 1,500 Oerlikon anti-aircraft guns for use on ships, a purchase worth about £5 million, on top of which came orders from the Air Ministry (although many of the items were not ready for shipment when the Germans launched their May 1940 western front attack). Needless to say, the British had also lodged purchases for various types of watches. But there were gaps: for one thing, as Wylie points out, the Italians allowed Swiss ships to continue using their ports. But then came smuggling, especially after 1941 when the Germans were able to close off Switzerland's parcel post shipments abroad: the first

items the British needed were the jewels used in instrument production for fighter and bomber aircraft. Wylie says that Latin American
diplomats, notably the head of the Panamanian mission in Berne,
helped in carrying out items that were then passed on to the British.
The Mexican and Venezuelan consulates were also part of the game.

# X

Germany may have invaded Denmark and controlled its great
agricultural industry, denying the British access to its products.
But they were not able to turn that into control of three Atlantic
territories which had been closely connected with Copenhagen,
another example of how Germany was hemmed it at the extremities
of its conquered territory. It could exploit everything within those
extremities, but little beyond.

Just days after the Danish surrender Britain set in motion plans
to occupy thsee territories. Troops landed on the Faroe Islands
(a Danish dependency) and Iceland, which was in a loose union with
Copenhagen. The authorities in Greenland refused to co-operate
and it was not until 1941 that Greenland saw the writing on the wall
and agreed to American forces landing there.

Any land base in the Atlantic held by Germany would have been
catastrophic for Britain whose merchant shipping was under constant
attack by German U-boats.

The Nazis were aware well before the war began that Greenland
was important, not just as a potential naval base but a more immediate
use—as a weather forecasting base for Europe. This was eventually
borne out when advance weather forecasts from German radio operators in Greenland of fog crossing the Atlantic allowed the Germans
to have the battleships *Scharnhost* and *Gneisenau*, along with the heavy
cruiser *Prinz Eugen*, risk a dash through the English Channel on
11 February 1942 and get back to Germany after having been bottled
up and bombed in the harbour at Brest on the French Atlantic coast.
On a day-to-day basis, German radio stations were sending weather
details several times a day to Berlin, information vital to U-boat

operations in the Atlantic. It was the fear of both the wireless bases and the potential for military ones that preoccupied the Western powers from the outset of war. Eventually, these weather stations were discovered and destroyed; in September 1941, for example, a vessel of the Greenland Patrol operated by the U.S. Coast Guard spotted a trawler flying Norwegian colours. The crew admitted to having just dropped a party at a remote bay—it turned out to be three Germans with a powerful radio transmitter setting up a weather station. Weather information became vital in another eventuality: after D-Day, with Allied armies pushing toward the frontier of the Reich, the Germans needed weather forecasts to have warning of bad weather, and therefore be able to move their ground forces without Allied aircraft able to find them through the cloud cover.

Greenland was vital in another respect: under Lend-Lease, the Americans needed to move large numbers of aircraft coming off their assembly lines to Britain. The best way to do that was to fly them across, and an airfield the Americans would build in Greenland would be a critical staging post on the circle route from Nova Scotia to Scotland.

Once the Danes surrendered on 9 April 1940, Washington saw the urgency of getting a foothold in Greenland. The following month the Coast Guard ship *Comanche* landed the new consul and vice-consul at Godthaab, the two men taking over the Danish doctor's house for their consulate. The Americans were reminded of the German threat with the sinking by a U-boat on 4 September 1941 of the destroyer *USS Greer* in Greenland waters.

The Germans did not give up trying. In August 1942 United States aircraft bombed a German wireless station on Sabine Island. German High Command in 1944 sent three more expeditions to Greenland to establish weather stations. On 1 September 1944 the cutter *Northland* intercepted the trawler *Coburg*, took off the twenty-eight Germans aboard and sank the vessel. A landing party from the *USCGS Eastwind* found a twelve-man German base. Coast Guard seaplanes located the third party which was trapped in ice near Cape Borgen aboard the white-painted German naval auxiliary

*Externsteine.* Again the *Eastwind* was in action, this time breaking through five miles of ice to get within firing range of the German vessel. Another U.S. ship, *Southwind*, was crunching through the ice from another direction and, when close enough, was able to catch the Germans in her searchlights allowing the gun-crews on *Eastwind* a clear view, upon which the Coast Guard vessel fired her five-inch shells to land near the *Externsteine* and scare the Germans into surrendering. It worked, with the twenty Germans giving up but able to destroy all their code books by the time the Americans came aboard. The Americans, on finding scuttling charges in place, made the three German officers remain on board their vessel to discourage their setting off the charges. The auxiliary was eventually freed from the ice and sailed to Boston where she became the *USS Callao*.

The British press initially did its best to portray the occupations of Iceland and the Faroe Islands in a rather rose-tinted hue. 'British Garrison's Friendly Occupation of Iceland', headlined *The Times* as the paper's Reykjavik correspondent reported on the first three months of the occupation.

The Icelanders, he reported, regretted the loss of their independence but acknowledged the necessity of the British move. On 10 May when the British had landed, no work was done in Reykjavik, the residents turning out to watch the troops in the harbour. 'The children run to greet any soldier they see, even tiny toddlers will leave their play and thrust their hands confidently into those of passing soldiers and walk along with them,' the report continued. But some undercurrents were also hinted at. The correspondent noted that Germany had always shown a great interest in Icelandic culture—and, by implication, the British had not—and Icelandic scholars and scientists had long been treated with great hospitality in Germany. Indeed, the majority of scientists and doctors in this nation of 117,000 inhabitants had received at least a part of their training in Germany, while German universities had devoted a great deal of work and money to the study of Icelandic language and literature. 'This interest naturally evoked a sympathetic response here, especially among members of the learned professions,' the article went on.

The Germans before the war had also sent instructors in gliding to work with the Icelandic aero club, sent teachers of skiing and rock-climbing, and invited football and athletic teams to the Reich. The reporter also hinted that some Nazi sympathisers were still around, and spreading anti-British rumours.

As far as the Faroe Islands were concerned, on 18 May 1940 the British photo-journalism weekly *Picture Post*—rather quickly off the mark and almost certainly having needed government co-operation to do the story—had a spread on the Faroe Islands. It was in the 'simple, sturdy islanders' mould of journalism with only passing reference to the British occupation, although it did make the point that the 3,500 inhabitants spread over seventeen islands had been dependent on supplies from the outside world, the reliability of which had been disrupted by the menace of German submarines.

Meanwhile, a 570 year old treaty brought part of the Portuguese empire into the war. Back in 1373, at St Paul's Cathedral, Edward III of England and King Ferdinand and Queen Eleanor of Portugal had signed a treaty of peace, with the first article stating the two parties 'shall henceforth reciprocally be friends to friends and enemies to enemies and shall assist, maintain and uphold each other mutually by sea and by land against all men that may live or die'. It this that Churchill quoted in a statement announcing Portugal's agreement to the Allies being able to use the Azores as a base. In October 1943 British warships sailed into the Portuguese territory's harbours.

One of the primary objectives was to have naval bases with which to persecute the war against the U-boats. Moreover, there was a feeling of great urgency about the need to improve the protection of merchant ships supplying Britain. By March 1943, the Germans had 250 U-boats at sea, and the central Atlantic was beyond the range of aircraft flying either from Britain or Newfoundland, and aircraft were proving to be a very potent weapon in the destruction of German submarine capability. One of the first tasks undertaken by the British was to bring ashore 60,000 American-made Marston plates, which were steel plates 3.05 metres long by thirty-nine centimetres wide. This gave them an all-weather 1,525 metre runway, allowing Allied heavy bombers to operate out of the Lajes field. By the end of October 1943, thirty B-17s and nine Hudson bombers were operating from the Azores. The first success occurred on 9 November 1943 when a RAF-flown B-17 from 200 Squadron attacked a German submarine. The airfield on the Azores was also used to transport new American military aircraft being flown to Britain, offering an alternative to the route via Newfoundland and Greenland

(The Portuguese had, as early as 1941, taken some precautions to prevent the Germans seizing the Azores. They lengthened the runway, posted more troops and stationed British-made Gloster Gladiators, a sturdy but slow biplane. Those aircraft were replaced in 1943 with American Curtiss P-36 Hawks.)

The Americans, who had thought about seizing the Azores back in 1941, put Pan American Airways up to starting talks with the Portuguese about using the airport on Terceira, also in the Azores, as a refuelling stop—the idea being to get the Portuguese to upgrade the airport. They also raised the possibility of a new airport on another island, Santa Maria. These talks were abandoned when the British began their negotiations with Lisbon.

Indeed, by this time, Churchill had proposed an occupation—by force if necessary—but was strongly opposed on this by the Foreign Secretary, Anthony Eden, and Deputy Prime Minister Clement Atlee. When the diplomatic alternative was initiated, the Portuguese were positive; after all, the Allied victories in North Africa had reduced considerably the ability of the Nazis to invade the Iberian peninsula. The Portuguese dictator, Antonio de Oliveira Salazar, was by this time convinced the Allies would win the war.

And, under the 'friends of friends' wording of the 1373 treaty, the Portuguese were able to approve American use of British bases on the Azores. The first American unit to arrive was the 96th Construction Battalion which was set to work developing a working harbour, unloading vessels and laying gasoline pipelines to the airfield. The Americans also had another card up their sleeve: they talked the Portuguese into allowing the construction of a second airfield on the

Azores with the promise that Portugal could participate in the liberation of East Timor, Salazar having expressed an interest in declaring
war on Japan (but at the right time—when the Japanese no longer
had the capability to seize Macau, of course).

**XI**

Business flourished in the most unlikely circumstances in occupied
Europe. And it takes a good deal to discourage speculation.

After the German occupation of France, the Paris stock exchange
was relocated to Lyons. By mid-1941, it seemed, the financial markets
were again active, the stock exchanges in France, the Netherlands
and Belgium having been closed for several months following the
occupation. A dispatch from Amsterdam in August noted that, in the
two years since the war began, the Brussels stock exchange average
was up 131.2 per cent, Paris-Lyons 80.1 per cent, Milan 75.9 per
cent, Berlin 65.8 per cent, Amsterdam by 24.2 per cent and Zurich
by 18.6 per cent; the London market, by contrast, was off 14.1 per
cent and Wall Street down by 8.3 per cent over the two years. The
Amsterdam market was trading strong in paper connected with business in the Netherlands Indies; just over three months before Pearl
Harbor, traders in Amsterdam were increasingly confident that a Far
East conflict could be avoided. They were buying stocks involved
with sugar, tin and rubber.

But there was a catch to this supposed investment bull market in
occupied Europe, and one that provides another illustration of how
Germany undermined the finances of its conquered territories, and
thus failing to harness these lands for the Nazi war effort. One more
bungle in a series of economic bungles.

The share market boom simply was not true.

A vibrant economy depends, in part, on freedom to do business. Investing in shares is very much an element of economic and
personal success. Not so in Germany and the German-occupied
countries. Initially the Nazis tried indirect means—stricter controls
of share trading, taxes on profits and capital gains and putting ceilings

on dividend payouts. But gradually more draconian controls were put in place: abolition of trading in futures contracts, the imposition of controls on the prices of individual leading shares, rationing of trading, and a ban on selling of treasury bonds. The rules were introduced in Berlin with immediate implementation on the Vienna and Prague markets, then extended to Amsterdam, Brussels, Paris, Milan, Bucharest and Budapest. Trading was reduced to negligible volumes.

Other than London, the only free market remaining was that in Zurich. More importantly, the apparent gains in the indices reflected not so much business confidence as evidence that traders were building inflationary expectations into the markets. The markets were also operating blind: there was scant information available about companies (no company information at all was being published in France). As *The Economist* pointed out in January 1941, the once great Paris bourse had 'fallen to the level of a small provincial market', with Germans forbidding dealings in industrial shares, while those in unoccupied France, at Marseilles and Lyon, had achieved unexpected performance. The Paris exchange was not even allowed to have dealings with its counterparts in Vichy France.

The Germans also set out to control postal and telegraph systems. Paul Ghali, of the afternoon newspaper *Chicago Daily News* foreign service reported that Zurich's German language newspaper *Neue Zuericher Zeitung* had disclosed Axis powers had, at a meeting in Vienna, signed an agreement to form a new European postal and telegraph union. Germany and Italy ordered its other members to be Albania, Bulgaria, Denmark, Finland, Croatia, the Netherlands, Norway, Romania, San Marino, Slovakia and Hungary. The union would be run from Vienna with the official languages to be German and Italian. Ghali surmised that one of the motives was to have the Reichsmark as the currency to be used in all postal transactions, along with imposing a common postal tariff through all the occupied territories.

The reporter noted that Franco's Spain failed to send a delegation, nor was there were representatives of the Vichy government. 'Switzerland sent two observers,' he added. 'But recent comments in the Swiss press leave no doubt of the suspicion and mistrust with which this country views the Nazi scheme.'

## XII

Logistics was a term unknown, or at least unused in the 1940s for other than military purposes, but in the transport sense of the term, Germany had more than its share of logistics nightmares. The Germans tried to impose control from Berlin on rail traffic throughout occupied Europe. All rail services were to be made subservient to the German war effort. But, along with the immense cost of the military machine and war materiel, the Germans found themselves having to build new roads and railway lines in places like Poland and Norway and, of course, the conquered parts of the Soviet Union. One such new railway line linked the Crimea via the Perekop Isthmus to Kherson in the Ukraine; this line was constructed after some of the bloodiest battles on the Eastern Front, the Germans taking the isthmus but losing large numbers of men in the process. New lines were needed elsewhere, such as Bulgaria; then, in addition to building new lines, there was the cost of double-tracking existing routes. Once within the borders of the Soviet Union, German engineering battalions had to convert the captured rail lines from Russian gauge (1,520 millimetres) to European standard gauge (1,435 millimetres).

Some lines in France were electrified to save coal. The French and Belgian railways were partly denuded as some of their rolling stock was transferred to eastern Europe where the bulk of the fighting was occurring.

To the deterioration in rolling stock through lack of repairs and maintenance was added the losses from air raids with British and American bombers frequently targeting railway depots and marshalling yards. According to the League of Nations, French State Railways, because of German seizures, damage and destruction by bombing, had by September 1943 lost the use of about thirty per cent of its locomotive stock, fifty-three per cent of its freight wagons and thirty-six per cent of passenger carriages. Belgian railways by this time had only one-third of its freight wagons still in service. The

Germans even ripped up rails in Belgium to be used for desperately needed new lines in the east. By 1942, 598 kilometres of track had been lifted in Belgium and removed.

And, just when Germany needed as many munitions, aircraft, tanks and other equipment, plants that could have been turning these out (as the railway workshops were in Australia, for example) had to step up production of new locomotives. Locomotive production in May 1943 was three times greater than it had been in 1941.

Meanwhile, the Russians were able to restore their own transport systems once the tide of war had turned. The victory at Stalingrad allowed the Soviets to restart shipping on the Volga (which before the war had carried half the U.S.S.R's inland water traffic). And while the Germans were dependent on their own industrial efforts or looting the occupied territories, the Russians had enormous help from the Americans who by the end of 1944 had supplied them with 362,000 motor vehicles, 1,045 railway locomotives and 478,000 tons of other railway equipment.

## XIII

While Germans, and the peoples of occupied Europe, were facing declining food availability, the turning back of the German armies returned to the Russians much of their own food producing areas. By the Spring planting of 1944, the Soviet Union was once again in control of all its major agricultural regions. Not that food was suddenly available as each area was regained: the Nazis had since late 1943 been transporting as much food as they could westwards, and destroying what they could not take with them. So the people of the Soviet Union were still growing hungry, too; but the loss to the Germans of these vast swathes of agricultural land was not only an economic blow but the defeat of the whole rationale of *lebensraum*.

### Postscript: The Beginning of the End

Winston Churchill on 9 November 1942 made, after the defeat of the German army in North Africa, his famous speech about beginnings

and ends. Of that moment in the war, he said: "Now this is not the
end. It is not even the beginning of the end. But it is, perhaps, the end
of the beginning."

For Germany in March-April, however, it was indeed very much
the beginning of the end. While this book is about the economic and
financial imperatives that, in retrospect, doomed the Axis from the
beginning, let us look at a few weeks at this period that, together, capture the plight in which Germany and Japan found themselves. Many
of the incidents retailed below will not be familiar to most readers;
there were far bigger events, more important victories, more dramatic
breakthroughs. But regard what follows as seasoning of the dish.

19 March 1944: The Admiralty announced that six German submarines had been destroyed in the North Atlantic by five sloops of
the Royal Navy's Second Escort Group. The ships, the *HMS Starling*,
*HMS Wild Goose*, *HMS Woodpecker*, *HMS Kite* and *HMS Magpie*, had
sunk the U-boats over twenty days of sailing to meet and then escort
a convoy, with three of the attacks taking place in one sixteen-hour
period. None of the merchant ships in the convoy was destroyed,
but *Woodpecker* was struck by a torpedo and, after eight days in tow,
foundered in heavy seas. All her crew members were rescued. The
Admiralty communiqué said the ships had made first contact with
the enemy 300 miles (482 kilometres) southwest of Ireland and the
submarine was depth-charged, the large pieces of the boat coming to
the surface after the explosions. 'Several days later the group joined
a homeward-bound convoy which was being threatened. Towards
nightfall *Wild Goose* sighted a U-boat diving one mile on the sloop's
port bow. As the *Starling* and *Woodpecker* closed at speed, the convoy
altered course to clear the area. Shortly afterwards, the periscope of
a U-boat broke surface only twenty yards abreast and to the port
of *Wild Goose's* bridge. The sloop engaged with gunfire, scoring
several hits until the enemy disappeared. *HMS Woodpecker* at once
attacked with depth-charges. Twenty minutes later deep underwater
explosions were heard and shortly afterwards star-shells fired by the
Starling illuminated much oil and wreckage on the surface of the
sea,' the Admiralty communiqué read. In the last kill of the voyage,

*Starling* engaged a U-boat with her four-inch and smaller armament, scoring several hits. The German crew of fifty-one abandoned ship and were made prisoners-of-war.

20 March 1944: Nazi troops occupied all of Hungary. London first heard about the German move over the Turkish radio station in Ankara. Berlin ordered the occupation after Berlin feared the government in Budapest would offer no resistance to the Soviet forces approaching its borders. Messages from Budapest, sent by a correspondent for Sweden's *Svenska Dagbladet*, stated that Germany invaded Hungary from the north and also from Romania. As London's *The Daily Telegraph* reported the next morning, 'the Russian advance to the borders of Romania has dramatically changed the situation. Within a few weeks, the Germans will be standing on the line of the Carpathian Mountains, which they must hold at all costs if they are to keep the Red Army out of the Danubian Plain, the gateway to Germany from the southeast'.

21 March 1944: Ankara Radio was again busy, reporting that Germany had made several moves to bolster its position in the Balkans. Several Nazi divisions were moving in Romania, while the Nazis had occupied post and telegraph offices in the Bulgarian capital, Sofia, and seized communications in that country. Formal occupation of Romania was deemed by reports appearing in British newspapers to be imminent. Meanwhile, all government buildings in Budapest were by this time reported to be in German hands and large numbers of Hungarians had been arrested, including many from the Foreign Office. In Russia, news reports were filtering through that German civilians were leaving Odessa in large numbers, the German garrison there being in immediate danger of being overwhelmed by the Soviet advances in southern Ukraine.

26 March 1944: The headline in *The Daily Telegraph* said it all. 'Russians reach Romania border on 53-mile front', said the first deck; the second deck added: 'Nazis back where they started 33 months ago'. After thirty-three months of fighting, the Germans had retreated more than 1,600 kilometres.

# 9.

# The Co-Prosperity Illusion

IT WAS CALLED MANCHURIA's boom town. While the Japanese wreaked plenty of havoc throughout much of China, they saw their capital in Manchuria as a showcase. The city chosen to be capital of the puppet state of Manchukuo was Hsinking (now Changchun). In 1937, an unnamed correspondent for the journal *Far Eastern Survey*, published by the New York-based Institute of Pacific Relations, described the effort the Japanese were putting into their new head-quarters. The location of the capital was chosen because that was the point at which the standard gauge (and Japanese-owned) South Manchuria Railway met the broad gauge (and Russian-owned) China Eastern Railway. Work had begun on the new capital in 1933 with planning for paved roads, water works, parks, sewerage systems, the idea being to create a city that could accommodate 500,000 people.

By the time the article was written, 6,000 new buildings had been completed, many of them houses to accommodate the influx of Japanese. By December 1935, there were 268,000 people living in Hsinking (including 51,700 Japanese). The article noted that the most impressive building in the city was the headquarters of Japan's Kwantung Army; at the same time, the puppet emperor Henry Pu-yi (by now styled Emperor K'ang Te) was still without a palace, instead living in the former headquarters building of the salt administration.

According to accounts at the time, the new Hsinking was a vast improvement on the old city where under Chinese rule 135,000 people lived; an American reporter described how, before the Japanese transformation, it had been 'one of Manchuria's shabbiest, most cramped and worst situated cities'. Now, instead of narrow lanes, there were wide boulevards, beneath which water and sewage pipes, along with telephone and electric cables, had been laid.

For all the show of their Chinese capital, the Japanese, far from harnessing the resources of its steadily growing empire across Asia, never managed to even get its original acquisition, Manchukuo, bedded down. This part of what had been Chinese Manchuria had been intended as home to one million Japanese farmers; only 300,000 arrived, and many of those returned to the home islands, unable to cope with the primitive conditions and the winter cold. Moreover, investment was slow: foreign capital could not be attracted and many of the large Japanese enterprises were kept out because of the Kwantung Army's hatred of the *zaibatsu*. The place was even less attractive as costs kept rising but wages did not; between 1938 and 1949, inflation in Hsinking pushed up prices by seventy-eight per cent. What commercial and industrial investment there was did not flow into benefits for the locals. Manchukuo was producing sixty per cent of the world's soybeans but food was short in the country. The borders with Chinese-controlled areas were porous. Hundreds of thousands of people crossed the frontier in both directions. The main beneficiaries, apart from the Japanese, were those who came from China to earn what was by their standards high wages, much of which was remitted home.

Hsinking and all its supposed splendours was just one part of what the Japanese conceived as the Greater East Asia Co-Prosperity Sphere.

The term started to emerge in Japanese publications with some regularity around 1940. According to a contemporary account, it was designed to give the Japanese people some concept of what they

were fighting for; secondly, the term 'co-prosperity' was intended as
a way to combat the enmity felt toward the invading Japanese by the
Chinese and others who were to become subjected to brutality and
exploitation, trying to beguile them instead with the fiction of a better
life. The actual boundaries of the sphere were never clearly specified:
initially, it was understood that Japan itself along with Manchuria and
the rest of China would be embraced, along with the Philippines, the
Netherlands Indies, French Indochina, Thailand, Malaya and Burma.
At some stage, the Soviet Far East was considered a potential part, and
by 1942 the Japanese had added India, Australia and New Zealand.
Incidentally, the Japanese were expecting to get their hands on the
large number of cars in those latter countries (808,500 in Australia,
276,000 in New Zealand and 123,000 in India) for use as 'smooth
transportation' in the co-prosperity sphere. The Co-Prosperity
Sphere was to remain an amorphous concept in everyone's eyes,
including the Japanese: at no stage were its boundaries defined nor
its purposes detailed although the mention of 'one billion people'
suggested that India was going to be part as well as China, while there
was also talk of the 'South Seas' (which no doubt where all the plans
for those motor cars in Australia and New Zealand came in). As it
was, the 'sphere' came to include 713.7 million people, including
Japan and all the occupied countries and parts of China.

In November 1942, the Ministry of Greater East Asia was formed
in Tokyo to oversee the administration of lands running from the
Amur River in Soviet Siberia to Tasmania in Australia; it was divided
into four divisions: General, Manchurian, China and Southern.

Japan stood alone in Asia as the only country to have progressed
towards industrialisation, and by early in the twentieth century its
economy had already been substantially transformed. In 1913, only
seventeen per cent of its imports consisted of finished goods, the
other eighty-three per cent consisting of either raw materials or half-
finished products; compare that with British India at the time, which
had finished goods making up eighty per cent of imports. However,
the figures do not tell the full story: Japan's exports consisted largely
of textiles rather than the complex mix of manufactures which

typified European economies. By 1937, though, Japan was a net exporter of machinery. And there was another sector in which it was making strides: canned food. A survey in 1939 by the Canned Foods Association of Japan showed the country to be the world's biggest exporter of canned fish, with forty per cent global share, and more than double the exports of the United States. (Overall, though, and taking all canned foods into account, Japan's production of thirteen million cases a year was dwarfed by America's 278 million cases.)

There is no question that Japan's industrialisation put it heads and shoulders above the rest of Asia. All well and good, though, just so long as the country could still get its hands on the raw materials with which to make those machines, especially iron ore.

The populations of these eventual parts of the short-lived Japanese empire had not been unimpressed with their later and temporary masters. Their manufactures had in the 1930s advertised the Japanese way of life. As the historian Charles A. Fisher noted crisply, 'it is difficult for the Westerner to realise the appeal that these often trumpery and shoddy goods exerted on the ordinary Asian peasant'. Those same peasants had grown resentful when, in the depths of the Great Depression, many of the colonial authorities restricted Japanese manufactures entering their territories to preserve what market was left for European goods.

The author of those words had further perspective into the Japanese methods as they captured European colony after European colony. Fisher, who would become a noted geographer, had been appointed to the Malaya Command in 1940 and was made a prisoner-of-war after the fall of Singapore, first at Changi prison and then on the Burma railway. As he was to note—with gentle restraint—in a paper on Japanese expansion into the Asian tropics:

> Europeans who, like the writer, were forced unwillingly to partici-
> pate in establishing the New Order were amazed at the Japanese
> genius for improvisation in this type of country, just as the world
> at large had been astonished at their military successes there after
> their unconvincing performance in continental China … Major

road and rail projects, relying on the most rudimentary techniques of wooden and bamboo construction for bridging and for the accommodation of labour were rushed through in less time than it would have taken European engineers to survey the route.

Japan in March 1940 revealed its plan to take commercial control of key resources in occupied China through newly formed puppet companies with a total value equivalent to $106 million, a key plank in its building of the cynically named Greater East Asia Co-Prosperity Sphere. One was the Central China Iron and Mining Company to control and exploit all mineral deposits. In fact, one of the first monopolies established in Manchukuo was gold: each area thought to contain the yellow metal was declared state mining property. Geologists were on the ground by 1935 and there was soon announced a large discovery in the far northern province Heilungkiang (now Heilongjiang).

The Nanking puppet government established by the Japanese (and led by a former Chiang Kai-shek associate, Wang Ching-wei) would not recognise any railway company other than the Central China Railway Company which would gain monopoly control of not only all the railway operations in China but motor services as well, taking over about 500 miles of rail track seized by Japanese forces in the lower Yangtze valley. The Central China Aviation Corporation would be capitalised with the equivalent of $6 million, with its head office in Peking. All communication links between Japan and China, and within Manchukuo and occupied China, would be in the hands of the Central China Telecommunications Corporation which would get into business with the yen equivalent of $15 million in capital. There was also to be a Central China Water and Electricity Company. All these companies were to be provided with tax exemptions, subsidies and land expropriation rights.

In late 1939, Japan, under the umbrella of the Japan North China Development Company, moved to oust foreign business operations from Chinese territory it occupied. It set up the North China Electrical Industry Company on the basis of taking control of existing power plants that had been confiscated from their owners. The country was gradually being turned into a subsidiary of Japan Inc: there was the North China Motion Picture Company and North China Cement Company among others. In May 1939 the Japanese founded the Huahsing Commercial Bank. Then, in August 1939, there was founded in Tokyo the East Asia Shipping Company which was a new umbrella for the Japan-China shipping operators including Nippon Yusen Kaisha, Yamashita Steamship and several other Japanese lines. The government in Tokyo decreed the new company would have a monopoly of all shipping between Japan and the ports of Tientsin, Tsingtao and Shanghai, as well as the bottoms trading between those ports and Formosa, various South China and North China ports including Dairen (also known as Port Arthur, and now Dalian).

Much has been written about Britain's naval blockade to starve Germany of vital supplies. Less well known is that Japan adopted the same tactic against China. As Britain was going to war in September 1939 against the Nazis, the Japanese were starting to apply pressure to British trade in the east. Ships flying the red ensign still had access to Hong Kong and Shanghai (where British registered ships accounted for about forty per cent of foreign trade) but those same ships had closed off to them many other ports along the central and southern coasts of China. In July 1939 Shanghai's outward shipping tonnage fell by twenty-seven per cent due to the fall in coastal shipping from other ports due to their being blockaded by the Japanese navy.

In May 1940, the Tokyo-based *East Asia Economic News* published by the Japan Economic Federation reported on plans for a canal to link the industrial cities of the Japanese-controlled zone—Fushun, Anshan and Mukden (now Shenyang)—with the seaport of Yingkou. Anshan produced eighty per cent of steel being made in Manchukuo, Fushun had a large coal mining industry and Mukden was an industrial powerhouse. The canal would extend ninety-six kilometres and connect with the Liao River, which would be dredged to navigable depth. Apart from the transport considerations, the proposal included reclaiming sufficient land along the river to accommodate 20,000 new farmers.

According to reports that surfaced in newspapers around the world—including the *Montreal Gazette*, from which account this is taken—in November 1940 Japan had announced a new ten-year plan for its incorporation of China into the Japanese empire. Japan would be the industrial power while Manchukuo and China would both feed in the raw materials for industry and then provide the markets for products of it. Manchukuo would provide sufficient staples for 100 million people of Japan, the thirty-five million in Manchukuo and 450 million in China. Chinese labour would provide the manpower to develop mines and other resources.

The report from Shanghai correspondent of *The Economist* in London concerning the Japanese army's control of the business and financial life of China took nearly two months from its dispatch from the Chinese city until it made into print, an indicator in itself how, by early 1941, the reliable flow of information faced roadblocks. Writing on 5 February, and published in the 29 March edition, the correspondent noted that Japanese's economic stranglehold was tightening in the occupied areas.

> The pre-war Japanese manufacturing industries in occupied China, and the Chinese enterprises which these industries expropriated, are all now strictly regulated by the Japanese military machine. Outside Tientsin and Shanghai, these industries have to sell their products within the limits of product quotas and maximum prices fixed by the military.

It was widely reported in January 1940 that Japan was planning to use its base in China to dominate civil aviation in East Asia. It avoided using British Hong Kong, and instead decided upon Bangkok as the southern terminal of its international service; an air service was to begin from Shanghai, offering a connection from Tokyo, and then would fly south, stopping at Canton and then continuing on to the Thai capital.

The Nationalist Chinese authorities, as long as they were able, continued to operate passenger services, using Chungking as a hub,

operating to Hong Kong, Kunming, French Indochina, Chengtu
(now Chengdu) and Moscow. The *Christian Science Monitor* correspondent seemed nonchalant about the potential hazards involved.
'The only major risk about journeying in Chinese aircraft is that one
may be shot down,' he concluded his report. 'It is a risk virtually any
traveller in interior China gladly takes if he has the cash and can get
a booking. After all, one is likely to be shot or bombed anywhere in
China nowadays!'

Corporate Japan concentrated its investments in Manchuria,
Korea and northern China; just slightly over ten per cent of business
investment flowed to Southeast Asia, with the occupied Netherlands
Indies receiving the greatest attention of that latter portion. By
1943, Mitsubishi alone had 1,600 people working in the region: the
breadth of activity was extraordinary. This one company had trading
arms in the Philippines, Malaya, Dutch East Indies and Burma dealing in sugar, copra, hides, rice, oil seeds, rubber, lumber, maize, salt
and tobacco; then there were the agricultural activities from cotton
in the southern Celebes to vegetables in the Philippines and the
occupied Portuguese colony of Timor; forestry and lumbering was
being carried out in Burma, Malaya, Indochina and Dutch New
Guinea. Mitsubishi acquired manufacturing capacity for making
vegetable oil, alcohol, for shipbuilding, making engines for small
ships, rice polishing and cement. In Burma, Mitsubishi was in cotton
spinning and leather; in Java it made machinery, batteries, paint, paper
and beer; in the Philippines it ran a match factory. Many of the
businesses were initiated at the behest of the army or navy; in the
latter's interest, Mitsubishi cultivated cotton in southern Celebes and
Bali, although this operation was soon switched to food production
after farmers (with no experience of cotton) were unable to control
insect infestation; the navy was also pushing vegetable production
to supply its ships. Leather was a priority of the Japanese authorities
for footwear.

The factories owned by the Dutch Philips electrical firm in
Surabaya were handed over to Tokyo Shibaura Electrical Manufacturing (later known as Toshiba). Mitsui and, again, Mitsubishi engaged

in building wooden boats for army use, with thirty shipyards spread over Southeast Asia.

Thailand, or Siam, was a separate case. The country had in common with Japan the fact that it had never been colonised and Tokyo for a while attempted to make the Thais their allies. On 21 December 1941 Tokyo and Bangkok had signed the Thai-Japanese Offensive/Defensive Treaty of Alliance which foresaw joint military operations and close economic ties. A policy statement from Tokyo trumpeted the intention to uphold the honour of Thailand as an independent nation and that the Thais would be partners in the Co-Prosperity Sphere. But, as historian William L. Swan notes, 'there was to be no passive waiting for Thailand to come of its own accord into the new order'. The Thais would co-operate, and that was to be that. The Thais were to know their place in the greater Japanese scheme of things.

There were at least two million ethnic Chinese living in Thailand when Japan took effective control of the country, Japanese troops entering Thailand on 8 December 1941. While Thai authorities had up to this point been as accommodating as possible to Tokyo and then co-operated in the hope of lenient treatment, no such quarter was likely by the Japanese toward the Chinese, the community which had staged anti-Japanese boycotts in the late 1930s in protest at events on the Chinese mainland.

## II

By the early 1930s, Japan's immediate fishing grounds—primarily, but not exclusively, in the East China Sea—had been showing signs of depletion. Once their forces were bogged down in China, and rice supplies were tight, the authorities in Tokyo decided to expand fisheries and boat-owners were pointed to southerly waters, especially the Taiwan Strait and the South China Sea. As Chen Ta-yuen wrote in a paper published by Murdoch University in Perth, this fisheries policy had dual purposes: one, it would allow control of the fishing grounds of Southeast Asia as one of the preparatory works of the

Greater East Asian Co-Prosperity Sphere; two, the exploitation of these seas would make a greater contribution to wartime food supplies on the home islands.

Before the Japanese army was able to conquer the countries of Southeast Asia, however, many of the ports had not been welcoming to Japanese fishing vessels as anti-Tokyo sentiment rose. The Dutch authorities promulgated new fisheries regulations aimed at controlling Japanese engaged in the industry in the Netherlands Indies and, in 1937, foreign fishing boats were prohibited from operating in Dutch waters. Similarly, offshore fishing in Malayan waters had also been dominated by Japanese who had migrated to the British territory, and the British authorities were becoming increasingly concerned about the security issues involved in Japanese vessels using their ports. Also in 1937, the Malayan authorities began tightening regulations to discourage the fishing fleet. But since 1933 the Philippines had ceased issuing licences to foreign fishing boats and Manila from 1937 started to become increasingly worried about Japanese intentions and the following year followed the lead of others in making it harder for the Japanese boats to fish around the country.

Many Japanese companies, faced with this hostility, chose to move their bases to the southern Taiwanese port of Takao (now Kaohsiung). By 1941, twenty-eight Japanese pair-trawlers were based in Takao, mostly operating in the Taiwan Strait as well as near Hong Kong and in the Gulf of Tonkin. Another twenty-three trawlers were under construction.

The southern Chinese island of Hainan also figured prominently in Japanese fisheries plans. Once the island was captured, trawlers were based there and canneries built to process both meat and fish. The occupation forces also established small industries needed to provide everyday necessities: there were factories producing glass, ice, leather and matches; brick and tile making was also stepped up. The newly created Hainan Electricity Company in 1943 revealed a five-year plan to build four dams for the generation of power, only one being completed by the end of the war. So great was Japanese investment on the island that it was administered by more than four thousand Japanese officials and military men. The military yen was the currency used, and Mitsui Company handled all the island's external trade. Chinese workers were forcibly conscripted with more than 10,000 locals employed across the railway, mining and electricity sectors. The army also stripped the island of food and livestock, it being estimated that a quarter of a million water buffalo were lost to farmers.

### III

The Netherlands Indies (now Indonesia) were, upon the German conquest of Holland and before the arrival of the Japanese, left to their own devices. The Governor-General, Jonkheer Tjarda van Starkenborgh Stachouwer, was henceforth able to govern completely under his own authority, although one commentator, Rupert Emerson writing in *Foreign Affairs* two months after Holland had been subdued by the Nazis, wondered whether the colony could survive as a Dutch possession. It would depend, he thought, on the economic importance of the islands and the policy taken by the United States should anyone threaten the territory. Upon the surrender of the Netherlands to Germany, Japanese foreign minister Hachiro Arita issued a statement that Japan would not permit the Indies to change hands, presumably because Tokyo had its own plans for the group which, with a land area of 1.9 million square kilometres, and was home to about seventy million people.

The Netherlands Indies, it was said at the time, constituted the richest colonial plum in the world. It produced rubber (thirty-five per cent of the world's total), tea, petroleum (twenty per cent of the world output, reaching 61.5 million barrels in 1939), sugar, coffee, tin, tobacco, copra, palm oil and many other commodities. The Dutch territory lacked only bountiful quantities of coking coal, iron ore and cotton. Since the Great Depression, the Dutch had made efforts to diversify the economy away from almost total reliance on agriculture; industries were being established to process metals and oil. Britain and France had their own colonial empires to provide many of these commodities, but the United States and Germany did not; by 1939, many Germans had settled in the Indies and some even joined the Dutch administration and military there.

Then there was Japan, which relied heavily on oil from the Dutch territory—and would rely even more heavily if it could not get oil from fields controlled by British and American interests. In reverse, the Indies were a market for Japanese cheap textiles that were being kept out of the more protected markets of Europe and North America.

In late 1940, James Bassett of the *Los Angeles Times* flew from Darwin in Australia, first to Koepang (now Kupang) on the Dutch half of Timor—'miserable main city of Timor on the poorest island of the whole attenuated archipelago', as he described it—and then to cities on Java and Sumatra. Japan, he found, was on every Dutchman's mind. The colonial government had placed an order for trainer aircraft with Ryan Aeronautical Company of San Diego but he was not sanguine about the territory's ability to defend itself. His dispatch, published on 24 September 1940, continued:

> In all, the Dutch probably could muster about 250 army (land-based) aircraft, plus another 150 seaplanes (of which the best are the tri-motored Dornier patrol bombers bought—oddly—from Germany last March. On order they have 360 American machines many of which can't be delivered for at least 18 months … Dorniers used by the all-Dutch crews thunder into the outer islands —remaining away for as long as three months. Scores of sheltered, secret refuelling stops dot archipelago coves.

For all its need for petroleum and raw materials, the problem was that from 1942 Japan could not absorb all that the Netherlands Indies produced, hence the collapse of that colony's economy due to that factor (and the absence of any market other than Japan) and the general depredations of war. Japan's inability to absorb all the product from the Dutch territory became more critical once the Japanese merchant fleet was attacked and sunk due to the attacks by the United States Navy and Australian aircraft.

The Dutch themselves could not be expected to have put up more than a delaying level of defence in the event of a Japanese attack: the army consisted in 1940 of only 50,000 men (four-fifths of whom were locally recruited Javanese and others) while the naval forces consisted of five cruisers, twelve destroyers, eighteen submarines and a number of miscellaneous vessels.

## IV

'Japanese Airline Alarms Australia' headlined *The New York Times*. 'No Commercial Justification Apparent for Service to Portuguese Colony' stated the second deck of the headline. Australians had become alarmed at news that Japan planned to operate a commercial airline service to Dili in Portuguese Timor (the eastern half of the island, the other being part of the Netherlands Indies). Dili was just 725 kilometres from Darwin. Just weeks before Pearl Harbor, the *Times* warned that this bold move as taking place '1,200 miles farther into the South Seas than Japan has ever penetrated before'. There seemed no apparent commercial justification for a daily flight by Dai Nippon Airways flying boats between Palau—part of the South Sea Mandate administered by Japan on behalf of the League of Nations— and Portuguese Timor. These trial flights had been made possible because Lisbon feared Japanese forces in southern China would invade Macau, and they saw the Timor concessions as a means of protecting their more important colony on the coast of China. (A year earlier, Dai Nippon had begun a Yokohama-Saipan-Palau-Bangkok air service.)

As Japanese historian Kenichi Goto notes, Japan had largely ignored Portuguese Timor until Tokyo left the League of Nations in 1933. Tokyo, and especially the Japanese navy, saw it as both strategically placed close to the Netherlands Indies and potentially easy pickings, with Lisbon hardly in a position to defend it.

As Sydney historian Robert Lee explains, 'for the Japanese, Portuguese Timor represented an opportunity, a neutral colony where they could reasonably expect some hospitality in the midst of the Dutch possessions whose resources—above all, oil—they so desperately needed'. For the Allies, Timor was a threat, a potential Japanese base for military action against the Dutch.

The Australian airline Qantas received permission in 1939 to begin a Darwin-Dili service, but the Japanese objected: Tokyo was putting pressure on Lisbon to allow a Japanese flying boat service from Palau, but by 1940 the flying boat service from Australia to Batavia (Jakarta) was stopping at the Portuguese territory. As Lee outlines, both sides used their offices in Dili to spy on other's activities.

When war came, the neglect from Lisbon was all too apparent. As Lee writes, there were only 300 soldiers, of whom fifteen were Europeans.

> The native troops were ill-equipped and even paraded barefoot. All but forty soldiers were in Dili. In addition there was a small mounted frontier patrol of forty men based in Bonobaro [a small town near the Dutch border]. This modest force was armed with 500 old rifles, ten machine guns and nine Japanese 20 mm guns dating from 1890. There was one Hotchkiss 47 mm gun and also seventy Mannlicher rifles.

In 1941, the largest foreign community in East Timor was Japanese— all thirteen of them, six of those being Dai Nippon Airways officials. In October 1941, the Portuguese gave in to Japanese demands to allow them to begin an air service. Just weeks earlier, Lisbon had also given way on the subject of diplomatic presence, and a Japanese consulate had been opened in Dili. The British, meanwhile, were talking in London to the Portuguese and were suggesting a garrison being established on Timor. But Portugal was still fearful of losing Macau if it antagonised Tokyo. The Dutch, though, showed a deal more resolve: on 5 November 1941 the authorities in Batavia imposed a ban on exporting gasoline to Portuguese Timor in an attempt to impede Japanese aeroplanes being refuelled at Dili. The British saw through the Japanese plan, too; they were concerned about Japan having a foothold in Timor because of its strategic consequences for the Netherlands Indies and Australia once war began.

Then came war.

Japanese forces arrived in Portuguese Timor on 20 February 1942, landing on neutral soil and justifying it by noting that the arrival there of British, Australian and Dutch troops had already compromised that neutrality. Facing them were just 250 Australians; as the Australian War Memorial records, there was no direct confrontation, the outnumbered Australians acting as a guerrilla force. Help from the local people sustained this operation; the Timorese provided food, ponies and acted as guides. Many Timorese were executed by the Japanese; in all, about 60,000 died during the war, one of the highest per capita civilian casualty rates in the Second World War. As Portugal was a neutral, that country's colonial officials continued to run the administration, but this changed suddenly in October 1942 when all Portuguese nationals were interned.

From July 1942, the Japanese set out to eliminate the Australian forces, by which time the latter's numbers has reached 700. The Allied position, notwithstanding some air support from the U.S. air force, became untenable, and by February 1943 all Australian forces had been withdrawn.


## V


The Japanese copied the German clearing system so far as money issuance was concerned. The military yen was the currency for Japanese controlled areas of China and Indochina along with Burma, the Netherlands Indies and the Philippines; in Thailand, a central bank was founded and the baht linked to the yen at a new parity more favourable to Tokyo. In Shanghai, once they took control, the Japanese authorities burned any U.S. dollar or British sterling notes they could find, abolished the gold peg and also any exchange rates that had previously existed with the dollar or pound.

The military yen was quite distinct from the original yen, the circulation of which was prohibited outside Japan itself. The rate of exchange between the yen and the military yen varied and was set separately for each conquered territory. The Burmese rupee, the Java

*A Burmese 100 rupee note issued by the Japanese occupying forces.* Author's collection.

and Sumatra guilder and the Malay dollar were declared to have a common conversion rate against the military yen, regardless of the fact that the pre-existing rates had differed widely. In Thailand, a central bank was founded and the baht linked to the yen at a new parity more favourable to Tokyo.

For a short period after the occupation of Malaya, British-issued Straits Settlements currency remained legal tender but eventually Japanese currency became the main means of transactions. It was known as 'banana money' due to the ten dollar note carrying a picture of that fruit. In fact, the first two years of the war before the Japanese attack had been ones of considerable prosperity in Malaya. Rubber and tin exports soared, local people were employed for defence works, and large numbers of military arrived.

When it seemed certain the Japanese were coming, the British colonial authorities ensured that bonds were called in and destroyed to prevent their falling into enemy hands, jewels and stamp collections were handed over for transport to Australia, and un-issued currency was either burned or sent to India. Coins were dumped in the sea.

But the military yen was in circulation soon after the Japanese had taken control and their own banks were opened, led by the Yokohama Specia Bank, Bank of Taiwan and Japanese Kanan Bank. The existing post office savings bank was re-opened as the Nippon

Government Post Office Savings Bank. New taxes were also imposed
and anyone considered wealthy dragooned into making contributions
to the new regime. Lotteries which returned about one-third of the
money collected from tickets were another means of extracting cash.

By 1944, the Japanese, to help finance their own operations,
resorted to the printing presses, adding to the money supply the
equivalent of about four billion Malay dollars without any economic
activity to back it up. In 1945, more money was printed, so that the
amount in circulation was many multiples of that which applied at
the beginning of the occupation.

## VI

The National Library of Singapore commissioned a number of local
historians to compile accounts of life in the colony under Japanese
occupation, during which the British colony was renamed Syonan-To.

The Cathay cinema had been opened in 1939 and the tower,
which contained the hotel and restaurant of the same name, was
completed just before the Pacific war broke out. The Japanese
made it their broadcasting centre (the British Malaya Broadcasting
Corporation having operated from there) for transmitting Radio
Syonan programs. The restaurant became the Japanese officers'
dining room. Outside the building, there were human heads stuck
on poles; these were from beheaded looters and other victims of the
Japanese military.

Three amusement parks were in operation before the war providing cheap entertainment. New World had opened in the 1920s,
followed by Happy World (which catered especially to children)
and Gay World during the next decade. These parks were a cheap
and cheerful mix of cabaret, ronggeng (Javanese dance), bangsawan
(Malay opera), movies, gaming, sport, stunts, circus and shopping. The
Japanese turned all three parks into gambling farms, handing their
operation over to Chinese businessmen. Licence fees were paid and
revenue shared with the occupiers. Japanese troops were not allowed
into the gambling sections of the parks.

One of the most elegant places to stay while in Singapore these days is the Goodwood Park hotel, just off Orchard Road. It was opened in 1900 and built to provide a home for the Teutonia Club, an organisation which had existed in Singapore since 1856 to provide comfort to German residents and travellers. During the First World War, the club was seized as enemy property and converted into an electricity generating station. It was auctioned after the war and run by three brothers who provided a style of country house living for those such as the Prince of Wales. It attained its present name in 1929. Not surprising then that senior Japanese officers saw as just the place to set up home for the duration. After the Japanese surrender, the hotel was occupied by the Australian army and used for war crimes trials.

## VII

Would the Greater East Asia Co-Prosperity Sphere have ever worked? On one level, never; on another, probably not. On that first point, there was never to be any 'co-prosperity' to speak of; the Japanese had about as much interest in sharing their wealth derived from Asian territories as did the British, French and Portuguese colonists. It was always all about the Japanese. On the second point, Japanese economic planning for the war was so inadequate, the new rulers clearly did not know where to start once they had done the military bit. Even if they had been given a few more years to bed down their conquests, it is unlikely that they would have known what to do with their empire.

Nevertheless, neither proposition would be tested. The war was soon to go against Japan and it was a matter of extracting as much as possible from the conquered lands in order to sustain their military and naval efforts. Each war theatre became its own entity, living off the land as it were, with no longer any thought or care for the organic unity that had impelled the Co-Prosperity Sphere concept.

Even when they had a spell of peace, the Japanese did little. While the battle for Shanghai in 1937 has been described as the

Asian theatre's Stalingrad, by the time of its liberation in 1945 eight years had passed with the Japanese making no effort to bring 'co-prosperity' to what was then Asia's greatest city. As *The Economist* reported in September 1945, 'not a single new transportation facility, public utility, heavy or manufacturing industrial enterprise, or big building of any type, has been constructed here during the whole of the period (since 1937)'. Quite the opposite in fact: many factories had been closed down due to lack of materials or fuel (which has become a very old story by this time through the Japanese conquered territories; they really did not have a clue when it came to running their new, albeit temporary, empire). What machinery was still operating was pretty much on its last legs.

The Chungking government run by Chiang Kai-shek and his cronies managed a very powerful piece of currency debasement: by the end of the war, prices in Chinese dollars were 2,500 times what they had been in 1937. But the currency introduced by the Japanese in China was even more worthless by 1945; you could buy—if the notion possessed you to do such a thing—up to 250 of the Japanese puppet dollars for one almost worthless Chinese one.

So much for prosperity of any type under Japanese rule.

# 10.
# Japan Feels the Squeeze

IT WAS ONE OF the most curious aspects of World War II that Japan made so little effort to be equipped to fight a total war and, when it did realise the scope of its economic shortcomings, still did too little, too late: it began to gear its economy for total war only in 1942. It, among all the belligerent nations, took the least advantage of women in terms of the economic contribution they could make in the workforce; it was, again, only in early 1942 that the government proclaimed the formation of the Greater Japan Women's Association to mobilise the female population. But its heart was not in the task. Eventually more women were brought into civilian jobs to fill the acute labour shortages but, even in the dark days of 1944, the Tokyo government did almost nothing in this regard; the most draconian measure was a decree in October that year that froze in their places the women already working in war industries. A census in February 1944 showed that women made up forty-two per cent on the non-military workforce (including agriculture), with a heavy weighting to those aged between fifteen and twenty-four; take out those working on family farms and over twenty-four, and the women of Japan were still largely outside the war effort. When it came to factory work, still as late as 1944 seventy-six per cent of employees were men. The female Japanese labour force rose by less than ten per cent between 1940 and 1944.

And the policy was nothing if not contradictory. Too little effort was made to get women harnessed for war work; but when they were, they were often badly treated. Few concessions were made to the comfort of those who did take up factory roles. As was mentioned earlier, Hirako Nakamoto was sent to work in a factory with a group of fellow students aged between twelve and fourteen; her assignment was a Hiroshima textile mill that had been converted into an aircraft factory. On those weeks when she worked the shift from three in the afternoon until eleven o'clock, Nakamoto was required to sleep at the factory overnight and allowed to return home only when daylight came. That was understandable, but then there were the frustrations. There were not enough machines for all the female students drafted to this particular factory, yet the girls for which there was no machine and therefore no work were allowed neither to sit down nor read. No meat, vegetables or fish were available to any of these young women in factories; those all went to feed the military. There was very little rice and they had mainly a red-coloured grain that was normally used to feed horses and cattle.

'At the factory, on the night shift, after standing for hours, we were marched into a dining hall where we had our supper. Supper was a bowl of weak, hot broth, usually with one string of noodle in it and a few soybeans at the bottom,' Nakamoto recalled. It was cold and, at nine o'clock on the late shift, the girls would each be given a small cake made out of weeds; however hungry they were, they could not bear the taste.

Before the war, many Japanese women worked but it was mainly in agriculture and on the family farm. Small family commercial firms would also employ women as assistants but very few females found their way into the professions or the civil service. From November 1941 unmarried women between the ages of sixteen and twenty-five had to enrol under a national registration system but, unlike the men (up to the age of forty) similarly required to register, they were not drafted into factory work. The government made it clear it wanted women to maintain family life. Even the new Greater Japan Women's Association thought that handicrafts and home-based industries were the best means of employing women. And never was it considered suitable that a married woman should go out to work. In the latter years of the war the policy began to change, and Hirako Nakamoto's experience was possible.

One area, however, where women were critical was, of all things, in coal mining. Australian academic Matthew Allen, in a paper on Japanese women coalminers in the 1940s, shows that females had worked in these mines from the early days of the industry, although they usually toiled on the surface while the underground labour was done by men; in 1933 a law was passed specifically forbidding the use of female labour underground. However by 1940 the government quietly sanctioned the return of women below surface. By 1943, due to the demands in other parts of the economy (and more attractive jobs), Japanese coal mining was facing a severe labour shortage. Henceforth, the larger mining companies had women working at the coalface.

Women, though, did have a role to play in marshalling the country's savings.

The Japanese had followed with great attention the formation in 1916 of Britain's National War Savings Committee; this was replicated in 1924 on the home islands in the form of the Central Council to Encourage Diligence and Thrift, with the creation of secondary councils at municipal level. Group saving was encouraged. Women's organisations were harnessed to spread the savings message. This meant that, when total war began in 1937, Japan had an operating mechanism for promoting saving. They needed to: the invasion of China had served to cut off loans from British and American banks. This led the following year to the creation of the National Savings Promotion Bureau with the intent of coaxing money out of the people rather than forcing them to pay higher taxes, although taxes did rise slowly but steadily right through to 1945. As historian Sheldon Garon argues, both the Nazi (who relied heavily on debt financing for the war) and Japanese leaders feared that pushing taxes too high would turn the people against the military effort. For Japan, the decision to bomb Pearl Harbor would just add to the financial burden of fighting. It says something for the Japanese dedication that, even in 1945, when the government set a target of increased savings of sixty billion yen, the populace surpassed that target and added 67.4 billion yen to their savings accounts.

Nevertheless, draconian methods were frequently employed to ensure that money was saved. Villagers would be refused rations unless they put enough money aside; a birth of a child required a further deposit. So it was not surprising that resentment grew as the war took more and more bad turns. The savings drive also proved an opportunity for women's organisations to campaign against alcohol, tobacco and mistresses, all items that many Japanese men clearly regarded as necessities of life.

Studies of the black market in Japan show that the usual despair and necessity drove the supply and demand, but there was also an element of protest. By October 1941 more than 100,000 goods, services and materials were under some form of regulation throughout Japan. The control had begun in September 1939 with ninety items. By 1941 hundreds more items had been added. Fountain pens, rice, insurance premiums—you name it, the government controlled it. A squad, the economic police, was established; in just one month in 1939 they arrested 244,000 people for flouting the controls.

## II

Even when it came to the black market, Japan was showing the strain, apparent when one compares the experience there with the black market in Britain, still under siege from Nazi Germany.

In mid 1941, *Picture Post* sent Anne Scott-James (who would become mother of Max Hastings, one of the leading World War II historians) to investigate the local black market. Her report 'I Take a Look into the Black Market' observed that there appeared to be a glut in cosmetics. You could walk into sweet shops or lingerie shores, and walk along back-street market stalls, and women would have their choice. Manufacturers were entitled to only one-quarter of their pre-war production levels but women were buying more cosmetics than ever.

The one gap in the regulations allowed firms which had made less than one hundred pounds worth of cosmetics to carry on as before; they were not subject to the twenty-five per cent restriction. These small makers were largely chemist stores which, say, made up a small number of pots of various creams. Scott-James said plenty of chemists had been approached to sell their rights to manufacture, added to which many new entrants were making products in their own homes. 'They know nothing about formulae or hygienic manu-facture; they just stir up some sort of mixture on the kitchen stove, cool it, mould it, wrap it in paper—and hawk it round the shops,' she explained.

So, inconvenience and disappointment were probably the most the British experienced. Compare that with what happened in Japan. While British women openly shopped for cosmetics, so determined were the Japanese authorities to deter people consuming that middle-aged matrons were posted on street corners to stop and shame any person they believed to be too well-dressed. Even matches and coal could be obtained through the black market; munitions factories had to resort to sending agents to the countryside to buy food so the workers had enough to eat to keep toiling at their machines.

In 1943, according to a report in the *Yomiuri Hochi Shimbun*, 11,464 shops were closed in Tokyo. The closures were ordered under the National Mobilisation Law and, according to the newspaper, this had freed about fifteen thousand workers to be available for war industries, of whom about one-third had been allocated to munitions factories.

Statistics do not begin to convey the human suffering wrought within Japan by wartime savings campaigns. By 1944, Japanese households were saving an extraordinary 39.5 per cent of their disposable income. While the American and British governments financed their war efforts largely by means of taxation, the Japanese preferred to get the people to save and so provide money that could be loaned to the government. The slogan of the time was 'luxury is the enemy' although there was a delicious irony in the fact that the advertisement promoting the slogan was sponsored by Japan's largest confectionary maker; it seems that the 'luxury' classification applied also to the basics of life with most people having not enough to meet all the necessities of food, clothing and shelter.

And what to do with all that rice (at least initially) grown within the newly conquered Japanese empire? Japan itself was, in the late 1930s, and unlike with regard to most commodities, practically self-sufficient in this staple. In addition, by 1942 the Japanese controlled most of the rice which hitherto had been sold through international trading and exported by Burma, Thailand and Indochina. Apart from the fact that the new masters did not have the shipping capacity to move the rice by sea, there were no ready buyers within the Co-Prosperity Sphere. Korea and Formosa were largely self-sufficient, as was Manchukuo. The largest customers for Southeast Asian rice had been India and Ceylon, with other buyers in Europe and Africa. All these were now closed off.

To complement its planning incompetence, the Japanese government had bad luck, mainly in the form of a drought in 1939 that hit western Japan and all of Korea. Rice reserves were quickly exhausted. That year the Rice Distribution Control Law was introduced, creating the Japanese Rice Corporation. It prohibited speculation in rice, and dealers in the grain were required to obtain licences to carry on their businesses and farmers were required to sell the corporation all their rice apart from that needed for their own consumption. By 1945, the government would control 63.7 per cent of the rice available in Japan. Naturally, the government control proved to be inefficient, exacerbating the shortages which grew with each year of the war. However, even in 1941 the rice reserves in Japan amounted to just ten per cent of the annual harvest; there was not much margin for error, or interruption to supplies.

After the first flush of conquests in Southeast Asia, the rice paddies of that region were able to send millions of tonnes back to the home islands. But the Americans soon put paid to much of that traffic by sinking large numbers of merchant ships. By 1943, Southeast Asia was able to supply only 1.2 million tons of rice to Japan. The United States Navy also largely cut off food supplies from Korea and Formosa, with only 268,000 tons getting through to Japan.

By the time rice did begin running short, the plight of the merchant fleet was even worse and what shipping was still available had to be given to transporting priority war materials. This meant by 1943 rice rations were being cut in Japan. In 1944 Japan tried to make what rice was available go further by mixing with it other cereals. As fishing fleets had been ordered into war service, fish was hard to get, a serious blow to nutrition as fish had provided nearly three-quarters of protein in Japan (the constrained areas of agricultural land making it impossible to raise livestock); the authorities ordered soybeans as a fish replacement in the Japanese diet. The minimum age for male and female labour service was lowered to twelve years.

By the beginning of 1945, the supply of food in Japan had collapsed. The black market prices reflected this: the price for one and a half kilograms of rice on the black market had risen from three yen in December 1943 to thirty-five yen in July 1945. The consumer goods market had collapsed in 1944, so using the same two dates for comparison, a box of matches had risen on the black market from one yen to eighty yen (the official price remaining at 0.4 yen).

### III

In 1943 it began to dawn on the Japanese authorities in Tokyo that the lack of central planning for the war effort was now turning into chickens coming home to roost. All the effort had gone into the military planning for conquering large swathes of Asia, but little thinking was evident about how they were going to consolidate those gains from both strategic and economic points of view. In March that year a Cabinet Advisers' Council was established in Tokyo which was made up of seven magnates from the finance and industrial sectors. It is astonishing that it was not until as late as October 1943 that the Japanese government set up a munitions ministry; but even then, the result remained chaotic with the ministry responsible only for aircraft production. The control of ordnance and shipbuilding remained with the army and navy ministries respectively.

Nor had Japan faced up to the ongoing struggle for supremacy between the army and big business. The *zaibatsu*, the large family-controlled conglomerates led by Mitsui, Mitsubishi and Sumitomo, were in favour of the war insofar as it would allow their regional expansion but they wanted to run the economy, not be subject to decisions made within the military. To some extent, business won: in 1943 the Imperial Planning Board controlled by the army was abolished and replaced by the new munitions ministry.

That year, too, some attempt was made to order industrial priorities. (Again, the reader today must be aghast that almost two years of involvement in a global war had passed without what seem a rudimentary item of war planning.) Finally, it was deemed necessary for Japan to expand its iron and steel, coal, light metals, shipbuilding and aircraft industries. Orders went out to textile, glass, pottery and other industries that were by now seen as non-essential to turn over their plants to operators of the now priority industries. Any industry which could not be converted to war industry work would be closed. It was a bit late: by 1944, American bombing was starting to knock out large parts of Japanese manufacturing.

Even parts of the co-prosperity sphere were being written out of Japanese planning due mainly to the severe losses inflicted on the country's merchant shipping fleet. The conquered territories in Southeast Asia were now regarded as secondary to war priorities; instead, dependence grew on what was called the 'inner zone'— Japan, Korea, Manchuria and North China.

More steel production in northern China was ordered to alleviate the pressure on shipping now carrying all the iron ore and coking coal to Japan from Manchuria. Land reclamation schemes on the mainland were to increase cereal production there, but this was contradicted by a parallel policy in North China to switch food growing to cotton production to meet the desperate shortages of the latter being faced in Japanese textile mills. This, in turn, hit coal production in Japanese-occupied areas as Chinese workers deserted the mines to return to the countryside in search of food.

The shipping crisis meant that only a fraction of the bauxite, chrome and manganese being produced in Indochina and the Philippines was loaded on ships for Japan; in Malaya tin and rubber piled up in warehouses there rather than reaching Japanese factories. Food could not be moved from Thailand, which had large surpluses. Destruction of transport systems in Burma meant rice and livestock could not be used to supply food. Japanese destruction in the Philippines led to the collapse of sugar production in that country.

*While Japanese industry struggled to meet the country's needs for wartime equipment, the Americans turned their considerable industrial power into a sledgehammer. Here another tank comes off the assemble line at Chrysler's tank arsenal plant, located in Warren, Michigan. Production of tanks at Chrysler set another record in December 1942 when its plant completed more machines in that month than it had for the whole of 1941. The automaker was also manufacturing Bofors anti-aircraft guns, small calibre ammunition, marine tractors, gyro-compasses, fire-fighting equipment as well as Dodge trucks for the army* National Automotive History Collection, Detroit Public Library.

In late 1944, Tokyo started publicising its intention to revive the plan to have a rail corridor stretching from Korea to Burma. Dusting off old plans was a desperate move to overcome the loss of merchant shipping. The Japanese said the completed corridor would involve a seven-day journey from Tokyo to Rangoon; there would be ferry links from Japan to Fusan in Korea and over the Yangtze River. On a map, it did not look too challenging: there were existing railways over most of the route with, ostensibly, only a few gaps to be filled, mainly in southern China and in what is now Cambodia. There was only one major hitch to this seemingly simple solution: apart from the ferry services across the Tsushima Straits which were by this time in range of American bombers, and were often stormy, there was the rail gauge problem. Japanese freight wagons could not be ferried because, being built to the 3ft 6in gauge (1,067 mm) of Japanese lines, they could not be taken off at the other end because in China the gauge was 4ft 8½in (1,435 mm). Then once you got to Indochina, the gauge was one metre, or 3ft 3⅜in. There were other problems: most of the route was single track, which limited the number of trains being able to use it at any one time. And it was unlikely that Japanese steel mills would have been able to produce the many miles of new rails required for either double-tracking or building the many passing loops that would be required.

Coal in the Netherlands Indies was another issue. Most of the Dutch territory's coal came from Sumatra or Borneo (the latter now the Asian coal powerhouse of Kalimantan) but, of course, the largest consumer was Java. Due to the growing merchant shipping crisis within the new Japanese empire, no vessels could be diverted to collier duty within the Indies; as Japanese historian Shigero Sato points out, before the Japanese invasion, vessels of the Japanese merchant marine accounted for only seven per cent of the shipping space available in the Netherlands Indies, most of the business being carried in British or Dutch bottoms. The invaders commandeered some of the remaining coastal vessels to help move the coal to Java but this was inadequate: the shipping capacity simply could not move enough to keep Java's railways, power stations and factories operating. The only option was to start mining coal on Java itself. Altogether six mines were developed but the infrastructure was minimal; in the case of one mine, the operation was located forty kilometres from the nearest road. The Japanese, clearly not wanting to spend the time and money to build either a road or rail connection, resolved this problem by forcing local villagers to carry coal on their backs but, again, that provided inadequate supplies. So, while coal stockpiles rose in Sumatra and Borneo, the railway operations were cut back across Java because there was not enough coal to keep all the locomotives in service.

Impoverishment of conquered territories had its economic effect on Japan, too. Tokyo controlled some of the most economically developed parts of China yet trade between the two fell dramatically. By 1941 Japan was exporting less to northern China than it did before 1937, and certainly was able to import far less except for coal.

What a contrast in the United States. America's railways were by 1945 gearing up for a vast transportation task—the movement of the huge quantity of war supplies needed for the invasion of Japan. Over the previous twelve months, the U.S. railroad companies operating the main lines to the west spent $200 million on new equipment and upgrading facilities. The war years had seen the railroad operators lay more double trackage to cut down on delays caused by trains having to wait to pass on loops along single track routes; they had also introduced more centralised traffic control and enlarged marshalling yards, not to mention buying new locomotives and freight wagons. It was all necessary: for example, the Atchison, Topeka and Santa Fe Railway—which was handling about thirty-five per cent of all wartime traffic moving between Chicago and California—by 1945 had 206 diesel locomotives in service and was using almost exclusively diesel traction (instead of steam) on the 740 km section between Winslow, Arizona, and Barstow, California. A train arrived or departed Winslow every twelve minutes.

## IV

Yet the Japanese were being told as late as 1944 they were still winning the war. But then reality had to be acknowledged, and the people told of the impending threat. Schoolgirl Hiroko Nakamoto hated one of her follow factory girl workers who voiced the opinion they were not being told the truth. She considered reporting the girl; after all, no loyal Japanese would say such a thing. 'Thus it was a great shock to us when the newspapers and radios informed us that the war was moving closer to us and we Japanese must be prepared for a last fight on our own territory, on our own mainland,' she recalled for American writer Mildred Mastin Pace in 1970. Japan was the country of the gods—she was sure those gods would protect Japan against what she thought of as the 'cruel Americans'.

# 11.
# China—Japan's critical failure

IN THE CHINESE WARTIME capital of Chungking, life became very difficult for all those outside the government and business elites. Inflation was getting out of control as the Chiang Kai-shek government printed more and more money to finance its war effort. In 1937 the Central Bank of China had undertaken to support the value of the yuan at $US0.29. In 1942, it abandoned its by then much reduced official exchange rate of $US0.05. By 1944 the Chinese yuan (or dollar as it was also known) was worth just 0.2 per cent its 1937 value. The middle class and most civil servants were reduced to poverty and misery. They existed by selling their possessions, and resentment grew against the privileged few who had access to American money and goods.

Compared to the vast volumes of materiel Britain and the United States poured into their Russian ally, China was left largely to its own devices. Yes, there was the Burma Road and, before that, the Americans flew as much as they could from India. But, in the event, Chiang Kai-shek and his forces were largely self-reliant. They needed to be: the areas under Chiang's control, called Free China, had only 2,700 kilometres of railways, 32,200 kilometres of usable highways,

13,000 kilometres of navigable river routes and a very limited air service. It took two months for Nationalist troops to march from Sian (Xi'an) to Chungking by foot, which was the only transportation option available.

And yet, for all the economic chaos and shortages behind the Nationalist Chinese lines, still the Japanese could not prevail. Japan spent twelve years in conflict with China and, while it inflicted dreadful economic damage, it failed to bring China to surrender. And that economic damage was there for all to see. Nevertheless, as Rana Mitter points out in his recent history of the Sino-Japanese War, the transformation of Chungking into the country's new capital (after Nanking had been abandoned by Chiang) was remarkable. As he notes, all the other major belligerents until the end ran their war efforts from their existing capitals. Chungking was not even the capital of Szechuan (Sichuan) province; that was Cheng-tu, now Chengdu, some 255 kilometres from Chungking.

Of the four main Allies which remained in the fight throughout (the others being Britain, the United States and the U.S.S.R—only France of the defeated powers would be given a permanent seat at the United Nations Security Council along with the four which withstood Axis aggression for the duration), China was by far in the weakest economic state—a factor that makes Japan's failure to subdue the Chiang Kai-shek government all the more telling. As Mitter explains, even before Japan attacked China was not self-sufficient in food and depended on large quantities of imported rice. It is some testimony to the argument that Chiang's Nationalists were not entirely the incompetents and irresponsible lot they have been made out to be. They managed to keep going despite government revenues plunging after the war proper began in 1937, the greatest hole caused by the loss of control to Japan of the Chinese Maritime Customs and the considerable volume of import duties they collected (see below).

In North China, the pressure by 1940 was on flour as imports from Australia and the United States had ceased, and the cost of flour had risen by 300 per cent. William J. Bowen, a reporter of Tientsin's *North China Star*, filed a report to his former employer, the *Los Angeles*

*Times*, describing the conditions in that Chinese city, noted that 'a pound of caviar here is cheaper than a pound of butter; a tin of corned beef is an expensive luxury'.

By late 1942, Ruby Queen cigarettes had risen from eight Chinese cents a pack to twenty Chinese dollars (a pack of Camel would have set you back 200 Chinese dollars) while a ton of coal was 900 dollars and gasoline seventy Chinese dollars a gallon; at that stage, a Chinese dollar was worth around one American cent.

But that was only the beginning. In the latter part of 1944 prices in Chungking were reported to be at 700 times their pre-war figures (it would reach 2,500 times by the end of the war) so the report in *The New York Times* in September 1944 was clearly putting a gloss on the true situation when it came to following a trip by Mrs Kong Kung-pao to the local shops. 'Although it is done without ration books, the day's marketing for a Chungking housewife is just as complicated a process, and perhaps a little more arduous, than is the daily marketing for the American housewife,' the reporter began, although he did note that—at the end of the shopping excursion— 'you would have wondered how Mrs Kong, and her husband and her four children were going to fare on what looked like pretty slim rations'. Mrs Kong was hoping the family's clothes would last to the end of the war, because there was no money to buy any new ones.

There was no pork in the market that day and, after enquiring about prices for other meat, Mrs Kong decided that the family would have to do without. Nor could she afford the bamboo shoots, and decided against any fish also after finding out about the prices for that staple. She bought a few snails, some peas and beans, brine-soaked spinach and twenty eggs; all that would have to last for the next three days, and the Kong family would eating a good deal of rice.

A study of China's wartime finances published in 1965 shows that civil servants suffered badly. Between the beginning of the war and the end of 1943, the rises of income levels of government officials were ten times less than that of the cost of living; teachers' incomes rose by a fifth of the rate of inflation. The author of that study, Arthur Young, recounted the case of a family which had saved over the years to send their son to university. On his eighteenth birthday, that sum was sufficient to pay for the son's birthday cake.

Food supplies were just part of the problem of everyday life in the wartime Chinese capital. Life in Chungking meant having to make do with what was available. Commuters from the outer areas travelled by horse-drawn carts and what motor cars there were used various forms of propulsion: raw alcohol, gas generated from coal or charcoal, or very low octane petrol. Buses and lorries used charcoal gas.

Inflation made life even more harsh and difficult. One correspondent of the *Christian Science Monitor* was, in 1943, charged the equivalent of $75 for a blanket in a Chungking shop. But the reporter also noted there were some areas with no apparent shortages; there were plentiful supplies, for example, of fountain pens, shoes and watches. The same newspaper in early 1944 opened a dispatch with the words: 'You can buy as much food in Chungking shops as you want—if you can afford it'. The correspondent, Gunther Stein, noted the shops were well stocked, and only a few foodstuffs were rationed. The local people were paying (in U.S. dollar equivalent) forty cents for a pound of bread, forty-six cents for a pound of potatoes while a pound of good rice would set them back $1.12, and one pound of lard $2.25, while a pound of locally caught fish was charged at $3.60 a pound. In local money, that shopping basket equated to 313 Chinese dollars. That was about one-tenth of a government official's pay. At that rate, they would have no money for twenty days of each month. In fact, they would not even get through ten days with enough to spend on food. At the time Gunther was writing, electric power charges were about to rise by eighty per cent.

Of course, inflation was just bad—sometimes worse—in other cities still held by the Nationalist Chinese, but Chungking was where the foreign correspondents were. Not that the Japanese-controlled areas averted similar runaway inflation. Reporter Brooks Anderson filed to *The New York Times* on 16 February 1944 that prices were soaring in Shanghai where the occupying forces had imposed a thirty per cent retail tax and a forty per cent levy on luxury goods. While the local dollar in Free China could be traded on the black market at eighty to the U.S. dollar, for the puppet currency some 110 dollars were needed to buy one greenback. Surprisingly perhaps, there were plentiful supplies of many American goods in Shanghai but only the most wealthy could afford them. In U.S. dollar terms, a pound of American coffee was $7.20, eight small cans of condensed milk $32, a dozen large cans of American pineapples were $55, while to have a suit made from English woollen cloth cost over $220. Meanwhile, a ton of coal cost $135. A study by Parks Coble of how Chinese businessmen fared in Shanghai under the Japanese shows that, using a base of 100 for wholesale prices in the city in 1937, that by December 1941 the index had reached 1,560. But worse, far worse, was to come: by December 1943 the wholesale price index was at 214,000, a year later 249,000 but by December 1945 in the aftermath of Japanese defeat it had reached 8,520,000. Paper currency was worthless.

(Not everything was bleak in Shanghai in 1944: more than 100 coffee shops had opened over the previous year, the live opera and drama theatres were packed each night, and there were horse races each Saturday and Sunday.)

An intelligence study compiled in August 1945 at the Office of Strategic Services in Washington noted that all the many factors which had caused the price spiral in China—shortages of raw materials, inadequate transportation, and the printing of money—had been exacerbated by hoarding and speculation. It found that, between 1939 and 1944, Chinese commodity prices had increased at a rate of between eight per cent and ten per cent a month (or 300 per cent a year at the upper end). By the winter of 1944-45, the monthly spiral had reached twenty-five per cent a month. 'By May 1945, the purchasing power of Chinese National currency had sunk to two-thousandths of its 1937 value,' the investigation noted. By June 1945 buying an ounce of gold cost 35,000 Chinese dollars.

It was officials in lower ranks of government, soldiers, school teachers and clerical and factory workers, all on relatively fixed salaries, who suffered the most. By the end of 1944, their incomes in real terms were a tenth of pre-war values. Coolie labourers were to some extent insulated, as they received a large percentage of their wages in the form of rice. The toll was not just financial: as the report notes, it was widespread malnutrition and debilitation through the ranks of the Chinese army that prevented the troops offering serious resistance to the Japanese forces.

The post-war years saw even worse inflation. The last denomi- nated issue by Chiang Kai-shek's central bank was the 5 million yuan note of 1949; a year previously, the postal service had put on sale a stamp costing 200,000 yuan.

The Nationalists did try and curb inflation; but it was inexorable (about ten per cent or a little more each month) rather than spiralling in a very short space of time. That it did not accelerate totally out of control was an achievement of sorts on the part of the government in Chungking given the state of China under Japanese siege. And because transport links between different parts of Free China were patchy, the rate of inflation varied by region. Government revenue flows were impeded for the same reason: it was often impossible for officials to get to each area to collect the land tax which was needed to pay the soldiers and civil servants.

One strategy tried by the Nationalist government was to sell gold to the public which they could use to buy and sell food and goods, a device aimed at stabilising prices to the much more constant level of the gold price.

## II

American newspapers were sometimes clearly trying to put the best gloss on the dislocations to Chinese industrial output: the *Christian Science Monitor* on 27 June 1942 reported that, by the time Chiang Kai-Shek's government had moved to Chungking in 1938, more than eighty million tons of machinery across all industrial groups had been transported at average 1,500 miles. But it was often on a wing and a prayer with, for example, belting for machinery being fashioned from old tyre casings, discarded rails being melted down into engine mounts and copper coins being punched into bearings. To protect electric power generation from enemy bombing, some of Chungking's generating capacity was moved underground. The paper also reported the whole cotton mills, arsenals and machine shops were located in caves.

Chungking had been a backwater. Just as recently as 1927, there had not been a single wheeled vehicle in the city due to most paths and tracks being steep, these gradients being traversed by going up and down long sequences of stone steps. Sedan chairs were a common method of transportation. When the Nationalists retreated there in 1938 they found a city of just 400,000 people living within seven square miles, most of them poor and involved in grinding labour on farms. Szechuan province had only four per cent of China's electricity capacity (and would rise to only eight per cent by the end of the war). The province was deficient in paved roads, railways and steel mills.

There was initially but a handful of building contractors available to transform Chungking. But plenty arrived with the Kuomintang after fleeing from Shanghai and Nanking. A new skyline started to emerge, large premises were thrown up for the Central Bank of China and other financial institutions, many new shops, department stores, restaurants, beauty parlours and tea houses, most incorporating the words Shanghai, Nanking, Peking, Tsingtao and other cities where the owners had once lived. (In 1942 the Kuomintang ordered the closure of tea and coffee houses and ice cream parlours in Chungking as they were seen as distracting people from war work and also as locations for black marketeers.) Some fifty makeshift buildings were thrown up in just forty days for those who were re-establishing Nanking's Central University in Chungking. By 1940 two hotels had been opened, along with a press hostel. That year, too, saw the assembling of the China Philharmonic Orchestra under the baton on Wu Po-chau, who had just returned from Belgium. Concerts often combined Chinese classical music with some old staples as Beethoven's fifth symphony and Rossini's William Tell overture.

And in 1943 an English weekly tabloid—all the type hand-set— hit the streets, being the rebirth of the *Shanghai Evening Post & Mercury*. But, as early as 1939, the displaced *Hankow Herald* (now Hankou) was publishing in Chungking, although keeping its original masthead name as a morale booster.

It was, to be sure, not quite on the scale of how the Soviets rebuilt their wartime industries in the Urals, but the redoubt at Chungking was enough to give Free China the fighting chance it so desperately needed.

Some 155 factories had been relocated to Chungking by May 1939; in all, about 350 factories had been shifted from the coastal cities to various inland locations. Chungking's new arrivals—most of them from Shanghai—including mechanical engineering works, machine shops, electrical engineering plants, chemical industries and mining companies. These were the essential providers of war materiel; there were also some very large operations, including the ex-Shanghai Ta Hsing Iron and Steel Works which were reported in the Singapore press to be operating twenty-four hours a day. That company had managed to move much of its Shanghai machinery, but some things had to be built anew; the company reproduced with local materials its Danish electric furnace left behind on the coast; they also managed to bring 350 skilled workers to Chungking. The Hua Hsing Machine Works had, from its new Chungking base, by mid-1939 manufactured more than 100,000 machine guns. The Yu-Fong cotton mill mad been moved from Chengchow (now Zhengzhou) with 52,000 spindles and its own power plant. Shanghai's Mei Ya silk weaving factory was in full operation, getting its raw material from Szechuan.

In the middle of the war between Nationalist China and China, Rockwood Q. P. Chin described conditions under which cotton mills in the new Chinese capital were operating. Chin was attached to the Nankai Institute of Economics in Chungking (it had moved there after its campus was destroyed by Japanese bombing in 1937) and he recorded his observations in October 1942.

'For a year now, Chungking has had a respite from bombings,' he wrote. It was a blessing for the city's cotton mills: since 1932, he wrote, cotton mills in China had been a target for Japanese aeroplanes. The consequent widespread damage to the industry, with replacement equipment unobtainable, had caused a serious shortage of cotton goods. The invasion of Hong Kong in December 1941 cut off yet one more source of cotton and a port for machinery to be unloaded.

*Inside the engineering and foundry hall of the relocated Yungli Chemical Industries Ltd. The company was based in Nanking and owned China's largest synthetic ammonium plant. The business was shifted to Szechuan province behind Nationalist Chinese lines and re-established in the small town of Wutingchiao.* Needham Research Institute.

The strain on the surviving cotton mills in unoccupied China was exacerbated by having to supply clothing to both the army and the several million refugees who had fled before the Japanese armies. Some mills could be moved in advance of the invasion; two large mills between them packed 16,000 tons of materials in more than 21,000 boxes and carted them more than a thousand miles to Chungking, with others being located to Sian, Yunnan, Hunan and Kwangsi (now Guangxi).

But the situation for the cotton milling industry was diabolical: only about 'a dozen or so' mills operating in Free China with about 200,000 spindles and several hundred looms to supply a population of two hundred million people beyond the control of the Japanese. This was just one-tenth of pre-war Chinese capacity and one-twentieth of Chinese and foreign-owned mills. A few new spinning machines had, by 1942, been brought into China through Rangoon but much of the capacity dated from the 1920s and the mills were almost totally reliant on cotton growers in one province, Shensi (now Shaanxi).

In 1943, when the American-owned *Shanghai Evening Post and Mercury* re-opened in Chungking, the editor Randall Gould reported back to the Christian Science Monitor. The *Herald*, which had made its way to Chungking after fleeing Hankow, came to an agreement that the two newspapers would publish a joint weekly edition. He described the typesetting room: it was chiselled out of rock to make it bombproof, and was inhabited with Chinese linotype operators setting English character type which they could not understand. The type was then placed in the wooden formes—the frame that holds a page together when it is placed on the printing press —by another Chinese worker, who also could not read English, and then the press was started using a portable electric motor.

However, one problem with relocating industries outside the largest seaboard cities was the lack of an industrial workforce. Farm workers had to be trained to use machines, a process that further hampered China's already badly damaged manufacturing efforts. Typical was the case of one foundry in Chungking where the new workforce refused to believe that that scrap steel, no matter how finely broken up, would not melt at the same temperature as pig iron. It took several scrapings of unmelted metal from the bottom of the furnace to finally persuade them. There were also frequent strikes as the new factory workers demanded pay hikes, and which the Nationalist government frequently caved into rather than endure more disruptions.

The problem for those factories that had successfully managed to relocate from the coastal cities (now occupied by the Japanese) to the still unoccupied hinterland was that they were cut off from the ports, and therefore the rest of the world so far as acquisition of raw materials was concerned. In 1942 plants in Free China produced just 30,000 tonnes of iron and 10,000 tonnes of steel. Coal production reached only six million tonnes. On the other hand, Chinese enterprises were successful in making fuel (including industrial alcohol, diesel and motor spirits) from vegetable waste, totalling some nine million gallons in 1942. Gasoline and alcohol were mixed on a 50-50 basis.

The relocation was vital for China's survival, but as a report in the *Christian Science Monitor* in 1939 noted, 'the newly arrived younger people sigh for vanished luxuries'.

### III

It was equated as a defence measure of similar importance to the Great Wall. In a radio address over a station in Honolulu in November 1940, China's consul-general in that city, Mui King-chau, said both the Burma Road and the Great Wall were 'constructed by enormous manpower, fashioned with crude tools without the aid of modern machinery'. But he added quickly that the Burma Road was, unlike the great structure in the north, a symbol of modern China.

The road's role as lifeline to the Chungking government is well known. It was the main conduit for fuel and supplies from British territory to the beleaguered Kuomintang government in its battle to withstand the Japanese.

As Mui pointed out, the road was really not new. It followed the route of the Old Tribute Road, so named because it was the road taken by Burmese representatives carrying their tributes to the Chinese emperor. Marco Polo was said to have travelled its length.

Work rebuilding the road began in 1937. Within eight months, the 1,150 kilometre road from Kunming in China to Lashio in Burma (the railhead for the line to Rangoon) was open to trucks. Not a single piece of modern machinery was used in its rebuilding. That effort involved constructing more than 300 bridges and 2,000 culverts. Some parts of the road were 8,000 feet (2,438 metres) above sea level. Hundreds of thousands of people were involved in the construction work; rather than moving all the workers as the road was constructed, local inhabitants and their crude tools were recruited as the work progressed through various areas. 'Recruited' is probably misleading: it was forced labour to the extent that each village was required to provide a specified number of workmen for a month (and many of whom had never seen a motor vehicle).

Over this road, with its gorges, sharp corners and deep valleys, trucks carried into China gasoline, medical supplies, machine and

aircraft parts. For their return journeys, the lorries were loaded with
tung oil, tungsten, tin, antimony, tea, silk, bristles and hides.

As more and more Chinese ports were captured by the Japanese,
and then Indochina was subdued so cutting the rail route from
Kunming to Haiphong, the Burma Road's importance increased.
Even trade with Russia—the source of much important raw materials—had to go by sea to Rangoon after Japan captured Sian and so
broke the Chinese link by rail to Russia.

By 1939 supply by air was problematic too. *Time* magazine, it
its inimitable style, reported in late 1939 that 'there are only two
airlines in the world which cancel flights when the weather is too
good'. Both China-Eurasia Airline Corporation and China National
Aviation Corporation preferred to fly 'mostly at night, usually in
filthy weather' because that had (so far) made it impossible for the
Japanese to shoot them down. The article marked the inauguration
of the CNAC service between Chungking and Rangoon

At the birth of the twentieth century, the British had considered
building a railway from Burma to Kunming (then called Yunnanfu)
and a route was surveyed by a Colonel Davies. However, the Chinese
were not keen on the idea and the British were unsure about the
commercial potential of such a line. When it was clear all of a sudden
as a result of Japanese advances in the 1930s that some sort of new
route would be needed once the coast was cut off from the Chinese
government, building a railway could never be done in the necessary
time frame. A road was another matter.

## IV

The pressures of the war were taking their toll on Chinese publishers by 1939, and this was most obvious in Shanghai, the centre of
the Chinese publishing industry. In a report to *The New York Times*
published on 3 September 1939—the day that Europe went to war—
Shanghai literary figure Lin Yichin (he had translated Remarque's
*All Quiet on the Western Front* into Chinese) reported that there had
been a large decline in titles being published, although one of the

large houses, Commercial Press, was still managing to publish a book a day even though it had been forced to move to the interior. But the smaller houses did not have the capital needed to relocate outside Shanghai, and sold off their remaining stock. And the tastes at the time of the reading public? Lin reported that the West Shanghai Book Store, of 277 Yu-yuen Road, was promoting *How to Win Friends and Influence People*, *Mein Kampf*, *Gone with the Wind* and *Hope in America*, all of which local publishers had decided to print their own editions. There was also considerable pirating of English-language textbooks as well as *Reader's Digest*, a year's subscription costing the equivalent in Shanghai dollars of seventy cents American.

Among all the devastation inflicted by the Japanese invaders in China, perhaps the most senseless was the destruction of hundreds of libraries, and millions of books. By 1936, the year before the Japanese invasion of China proper, the country possessed 4,041 libraries; the largest, the National Library in Peking, possessed 617,000 books, more than a sixth of those in Western languages. The first library to be put to the torch by the Japanese was in Tientsin, followed by such libraries as the Hopeh Medical College and the Shanghai Municipal Library. The Sinological Library in Nanking was sacked (with much of the rest of that city), and its rare books and manuscripts destroyed. By 1939, 2,500 of the country's libraries had been lost.

By this time, however, American librarians had swung into action. Dr T. L. Yung, chairman of the Library Association of China, enlisted the help of the American Library Association to help supply colleges and campuses with Western reference works. The most urgent need was said to be for scientific and technical volumes. The Americans were given the example of Tsinghui, Peking and Nankai universities which had moved more than 1,600 kilometres to Kunming in Yunnan province, and combined to set up the National Southwest Associated Universities. Their combined libraries of 600,000 volumes had been reduced to just 50,000 books surviving the Japanese attacks. Public libraries also needed books. But Kwang Tsing Wu of the Library Association of China added one stipulation: 'I don't think we need any light reading', he told the *Christian Science Monitor*.

By the end of 1939, 24,000 books had been donated. Eventually the effort grew, the US government became involved, and the Books of China program was born.

V

There was one area of authority still intact after the Japanese occupied great swathes of Chinese territory: the Chinese Maritime Customs. Established in 1854 under Horatio Lay and best known for Lay's successor, Robert Hart, who ran the agency from 1863 until his death in 1911. Founded to establish a revenue gathering service after the Taiping rebellion, it went on to set up the Chinese post office, it provided pilots for river shipping, ran the coastal lighthouses, provided weather information and raised loans for the Chinese government.

The Chinese Maritime Customs resisted manfully efforts by Japanese forces to interfere with its operations. This was due largely to the service being run by British employees and it was only under the rule of Sir Frederick Maze which had begun in 1929 (he succeeded his uncle in the job) that Chinese were appointed to some commissioner's posts. The foreign employees of the service were the instigators of the rigour with which smuggling into China was controlled. When full-scale war between Japan and China broke out in 1937, the foreign staff stood their ground; by December 1941 and the entry of the United States into the war against Japan, only three Japanese commissioners had been appointed within the Chinese Maritime Customs, the remainder still being nationals of Western countries. With the advent of Pearl Harbor, of course, all British and American employees still at work in the Japanese occupied areas were dismissed, several being imprisoned for some time in Shanghai and subjected to the then common Japanese cruelty. Those who were left of the inspectorate were grouped around the headquarters established in Chungking. Again, the highest echelons were dominated by Europeans and offices established in interior locations to collect duties and taxes.

*Chungking, the wartime Nationalist China capital, posed severe logistical challenges. Here newly appointed United States ambassador Clarence Gauss is transported on 26 May 1941 to present his credentials to the Kuomintang government.* Association for Diplomatic Studies and Training.

Maze retired in 1943 due to ill health. By October that year, the acting inspector-general was an American, Lester Knox Little, the first of his nationality to hold the post. As correspondent Brooks Atkinson telegraphed to *The New York Times* in October 1943, each day Little went from his house on the hill in Chungking to his office by the Yangtze River. 'One round trip he makes on foot up and down the interminable steps that wind through a slatternly, noisy Chinese village. On the other trip he gives himself the mercy and luxury of a sedan chair,' he wrote. Little had been repatriated the year before after being interned in Canton by the Japanese. It was no easy job to establish a new trade and revenue system given that the Chinese government in Chungking was 'cut off from the coastal ports, shorn of her fleet, deprived of a large number of trained employees'. All those working for the Chinese Maritime Customs were living close to the bone; the hideous inflation in the areas controlled by the Kuomintang had eroded the value of salaries, even for the foreigners paid in sterling.

## VI

In 1944, the population of the island of Formosa, a Japanese colony gained in 1895 after the first Sino-Japanese war, stood at 6.7 million. Of the 450,000 Japanese who had lived on Formosa before the Pacific war, all but 2,000 technical and management people had been repatriated to the home islands. Japanese rule had seen considerable industrialisation: when the Chinese took over in 1945, they inherited the Japanese-built sugar refineries, a pulp and paper industry (using the 200,000 tons a year of bagasse from the sugar plantations), cement plants and fertiliser manufacturing. The water level of Sun Moon Lake had been lifted thirty-seven metres to fill a master dam, and total generating capacity in Formosa at the end of the war was 321 megawatts. Many of the forty-five sugar factories had been converted to producing alcohol. The pulp and industry's output was equivalent to about one-third of the entire output from the Chinese mainland. Three cement plants were in operation, the newest having been opened in 1943.

Eight months after V-J Day, *Time* magazine reported in the issue of 10 June 1946 that 'sugar-starved China was getting supplies from its new sugar bowl, Formosa'. Stores taken over from the Japanese army were being shipped to Shanghai. The people of Formosa, though, were not happy: the magazine reported the locals were angry at the Chinese occupation army looting stocks, and making no effort to keep crops growing or refineries, railways and power plants operating. The newly installed governor Chen Yi installed his nephew as head of the Taiwan Development Company, the entity used by the Japanese to control the island's industry and trade. *Time* summed up the local sentiment in its usual snappy style: 'Most foreign observers in Formosa agreed that if a referendum were taken today Formosans would vote for U.S. rule. Second choice—Japan'.

# 12.
# Losing the propaganda war, too

ONE OF THE VOICES heard on Japanese shortwave was that of an 80-year-old woman initially identified as Mrs Henry Topping. In August 1943 the Office of War Information released a report confirming that Mrs Topping (her full name being Helen Fayville Topping) was the voice behind the English language broadcast called 'The Women's Hour'. She and her husband (who had died in 1942, aged 86) had arrived in Japan as missionaries in 1895 and remained there after their retirement in the early 1930s. The American authorities said Mrs Topping was what they called 'an extreme pacifist'. Her programs were aimed at American women, and OWI included a passage from one broadcast. 'Can you tell me why the women of America should send their sons, their husbands, their brothers, to die miserably in the African desert or in the swamps of the South Pacific islands, or in the shark-infested waters around those islands', she said. OWI had also noted another broadcaster using the name 'Miss Francis Hopkins' who reassured her listeners about the 'peace-loving' nature of the Japanese, the *Washington Post* reported.

A powerful radio station operated by the Japanese was Radio Hsinking in Manchukuo. Located in the capital of their puppet state in North China, the station had opened in 1934. The 100 kilowatt transmitter, operated by the Japanese-owned Manchurian Telephone and Telegraph Company, was used to beam programming to Hawaii, North America, the Philippines, China, the South Pacific and Southeast Asia. Its daily broadcasts went out in Japanese, English, Mandarin and Russian.

On the morning of 7 December 1941, the day the Japanese attack on Pearl Harbor brought the United States into the war, *The New York Times* published one of its regular shortwave reports from W.T. Arms. A Japanese radio station based in Shanghai, he told the paper's readers, ended its daily English-language broadcast with the words 'V stands for victory, a German victory over the enemies of Europe'. The shortwave broadcast was aimed at listeners in North America.

Tokyo Rose and Lord Haw-Haw have become part of the folklore of radio during the Second World War, but there are many other stories that have been largely forgotten. And, as in so many aspects of the conflict, the Allies were often caught napping on the radio wars because they had not been paying attention to the growing potential of the medium.

The Italians were the first to use shortwave radio to cultivate support in areas they saw as future conquests, most notably using radio programming to soften opinion in the Middle East ahead of the planned invasion of Abyssinia. The Italians had a powerful medium-wave transmitting station at Bari, on the Adriatic Coast, which had been built to broadcast to their colony of Libya. In 1934 they added broadcasts of Arabic music and news, and these were relayed by the shortwave station near Rome. This meant Italian broadcasts were reaching large areas of the Middle East, including areas of British influence (Egypt, Palestine and the Red Sea region) and where French interests lay (Tunisia and Morocco). Initially, the broadcasts were not anti-British but were more concerned to promote Italy and the Fascist system. By 1935 there were more than 10,000 wireless licence holders in Palestine alone and the Italians sold subsidised radio sets in the region, particularly to café owners who saw a chance to provide entertainment to their customers. One British report, noting the combination in the broadcasts of propaganda and entertainment, commented that the Arabic speaking customers in these cafés 'sipped their coffee and swallowed Italian propaganda with every mouthful'. Italy began shortwave broadcasts to Latin America as early as 1933, with the Germans and Japanese following in 1935 (and the Germans also purchased local South American commercial stations, usually ones operating on shortwave to reach audiences in rural areas with no local medium-wave reception).

After 1938, the Axis broadcasters divided up the region: the Italian stations concentrated on attacking France in broadcasts intended for listeners in North Africa, while broadcasts from the German station at Zeesen, thirty kilometres south of Berlin, addressed mainly the Arabic speakers of the eastern Mediterranean. To these listeners, the Nazi programmes peddled a line about the imperialist intentions of Britain and France and told the audience the Allies wanted to rob and enslave Arabs. The Americans were linked with 'Zionists' in the broadcasts and, after the Free French took up the struggle on the Allies' side, General Charles De Gaulle was portrayed as an agent working to get control of North Africa for the communists in Moscow. The end game was to incite armed rebellion against the Allies. By 1938, the British government was concerned by the relentless hate being sent out from the station at Bari and the British Broadcasting Corporation was instructed to begin Arabic broadcasts.

As Germany conquered areas of the Balkans, it was able to also gain control of radio stations there. In addition to the Berlin transmitters, the Germans used radio stations at Bucharest, Sofia and Tirana (captured by Italy) to reach listeners in Turkey, ensuring good reception in that target area. Complementing the work being done by radio, Germany maintained six news agencies in Turkey and bulletins were sent out several times a day to newspapers across the country; for the smaller newspapers, there was the appeal that the Germans were more efficient than the domestic news agency and also supplied material free of charge. German agents also helped help compliant newspapers acquire scarce newsprint. There was a German language newspaper along with one in French operated on behalf of Vichy.

Just as the Nazis acquired radio stations as their armies brought more territory under Berlin's control, so Japanese conquests in Asia meant a vastly expanded broadcasting network available to Tokyo. By 1943 Japan controlled about forty broadcasting stations and was able to have programmes going out on shortwave around the clock on more than fifty frequencies. Radio Saigon was designated to carry anti-British broadcasts to India, backed up by similar programs emanating from transmitters in Singapore and Bangkok. Japanese forces captured intact a powerful medium wave station at Penang, Malaya. For local audiences, the Japanese also had at their disposal powerful medium wave stations in places such as Hong Kong, Manila, Rangoon and Batavia. In fact, the Japanese were able to reach so many more listeners than the Allies by using a combination of shortwave (for long distance) and medium wave (for local audiences), while the Allies broadcasting from Australia, the United States, India and Chungking had only shortwave, with all its unreliability of reception due to changing atmospheric conditions.

Tokyo was able to play on Asian resentment at the misdeeds of the expelled colonial powers. Radio Tokyo, in its Chinese broadcasts, devoted a week to the Opium Wars; listeners in India were reminded of the 1919 Amritsar Massacre, and were also told that Indian troops were placed in the front line in Malaya battle to take the brunt of the fighting.

Japanese stations from early 1942 sent out broadcasts in Arabic, Persian, Turkish and Pashto. The programmes portrayed Japan as champion of Muslims, and that it would help 'liberate' the Middle East, and the Muslims of India, as it had already done for 120 million Muslims living in the Philippines, Burma and the Netherlands Indies.

Shortwave during the Second World War was the main medium by which the warring nations could speak to the populations of their enemies, those of their allies and those of the neutral countries. While the Free French were bouncing shortwave signals off the ionosphere the Vichy interests were doing the same using powerful transmitters in Algeria and Morocco.

Competition was fierce between the two French sides, and they had on call a sophisticated shortwave network that had been built up in the colonies. There were two transmitters in Djibouti, station FO8AA in Tahiti, Radio Saigon, as well as stations based at Guadeloupe in the French West Indies, Madagascar and New Caledonia. Under the pro-Vichy authorities in Indochina, anti-Allied broadcasts emanated from the transmitters of Radio Saigon until the Japanese took control. Then the programming grew increasingly under Japanese influence with more languages added, broadcasts from Saigon going out daily in French, English, Mandarin, Cantonese, the Indochina languages and various Indian dialects. Bulletins in the Malay and Dutch languages ceased once Japanese forces had conquered Malaya and the Netherlands Indies. While the station was run by the French staff, the Japanese ran two daily English language broadcasts beamed to India and a tri-weekly session for Australia. One announcer in Saigon was a 36-year-old Englishwoman who had married a Frenchman. She was described in 1943 by London's *Sunday Dispatch* as 'Lady Haw-Haw'.

Meanwhile, in Africa, the Gold Coast broadcasting service was also to play its part in the struggle of the French-speaking airwaves. The British colony (now Ghana) was surrounded on three sides by French West Africa, initially under Vichy control and therefore hostile to the British cause. The British radio station run from Accra began a nightly half-hour transmission in French and some languages spoken in those territories (Ewe in French Togoland, for example). London, which had previously been parsimonious when it came to paying for radio stations in colonies, even bought a higher powered transmitter to be shipped to Accra.

In November 1940 Free French authorities inaugurated a new and powerful shortwave radio station in Brazzaville, French Congo (later known as French Equatorial Africa, now Republic of Congo). It was another voice in the battle of the airwaves. And by 1943 a 50,000 watt transmitter was operating in the Belgian Congo. It was used by the BBC to relay broadcasts to the United States, the signals from transmitters located in Britain to North America being diminished by severe fading due to magnetic interference over the polar path.

Then there was the American problem. Of all the main belligerents, the United States was alone in not having a government-run shortwave service when the war began. It was not until fifty-six days after Pearl Harbor that the Voice of America went to air (on February 1, 1942). Even then, it broadcast to Europe using BBC transmitters. The transmissions were in German, French, Italian and English. A year later, VOA had twenty-three of its own transmitters in operation and was programming in twenty-seven languages. This rose to more than forty languages in 1944.

It was estimated in 1939 that, of the eleven million radio sets owned in Germany, about five million had shortwave bands. More-over, the compact nature of Europe meant that, for those other six million wireless receivers, Allied broadcasters could reach them with powerful long and medium wave transmitters.

Reaching Japan, however, was altogether another matter. The sheer distance across the Pacific meant that only shortwave signals could span the gap. The Americans were also hampered by the fact that there were so few shortwave receivers in Japan because of Tokyo's policy that only medium-wave receivers could be sold, therefore placing a severe limitation on the potential audience for their Japanese language broadcasts. But in the Netherlands Indies, where shortwave had been widely used by Dutch settlers and officials to hear broadcasts from Holland, the Japanese either confiscated the sets or sealed them to just the Japanese wavelength.

In the United States, shortwave broadcasting had been in the hands of private broadcasting companies; and the two networks, National Broadcasting Corporation and Columbia Broadcasting System, operated shortwave services aimed only at South American audiences, but these were programmed as entertainment, financed by advertising revenue. Even then, NBC and CBS concentrated on the audiences in Rio de Janeiro and Buenos Aires, the main population centres, and largely ignored the regional areas and the entire west coast of South America.

The upside was that South America had a very substantial radio audience; there were more than one million sets owned in Argentina, 500,000 in Brazil, 300,000 in Mexico and 160,000 in Chile, with smaller numbers in other nations. Given the huge German propaganda and intelligence effort in Latin America, reaching radio listeners with the Allied message was vital. It had been reported in 1938 that Germany's radio stations were already beaming hours of programmes into the South American region using the Spanish and Portuguese languages. A *New York Times* report cited an hour-long broadcast to Bolivia on 26 November which included 'a very attractive musical program' after which there was a talk on how Bolivia could benefit from doing business with Germany as opposed to being exploited by American capitalists. The United States initially had no weapon with which to combat this radio warfare. In fact, it was the concern about Nazi broadcasts to Latin America that first began talk and consideration of a federally-operated overseas broadcaster.

Feedback indicated that, apart from news, South American listeners tuned in to NBC and CBS broadcasts to hear popular artists such as Rudy Vallée and follow news about American movies and Broadway gossip. In 1939 large numbers of listeners tuned in to hear the heavyweight boxing when Joe Louis met Chilean Arturo Godoy, South American heavyweight champion. Sponsored by Standard Oil of New Jersey, the NBC signal was re-broadcast over 130 local stations in Latin America.

NBC, using its powerful transmitters at Bound Brook, New Jersey, also began broadcasts to Europe in German for an hour starting at 8.00 pm Berlin time. The news bulletins contained items that were otherwise denied German listeners. According to *Time* magazine, in January 1939 these broadcasts included the news that Bridget Hitler, the Irish estranged wife of Adolf's brother, had been arrested in London for not paying her rent—she moved to the United States later that year and eventually changed her name—and that Washington viewed with alarm the dismissal by Hitler of Hjalmar Schacht as president of the Reichsbank.

It was not before time that America's voice was heard loudly.

In the months after Pearl Harbor, the privately owned shortwave transmitters were leased by the government although NBC and CBS still produced programs in Spanish and Portuguese in their studios for transmission southwards. By 1942, *Time* was reporting on La Cadena de las Americas (the Network of the Americas) describing how shortwave station WCBX was transmitting in Portuguese to Brazil while, from an adjoining studio, a Spanish-speaking announcer was reading the news to go out over stations WCRC and WCDA. The network had been launched the precious week at a swanky dinner in New York attended by Edward G. Robinson, Rita Hayworth, Ronald Colman and other movie stars, along with the presidents of Peru, Venezuela, and Nicaragua. Vice President Henry A. Wallace addressed the room—and those listening on their radios through Latin America—in Spanish. The CBS broadcasts were relayed over seventy-six domestic radio stations in South America, each station contracted to relay for at least one hour a day. NBC had more than one hundred local stations, including thirty-one in Mexico, relaying its broadcasts. By this time, too, other private companies had begun shortwave broadcasts, mainly to Europe: General Electric operated stations WGEO and WGEA, the World Wide Broadcasting Foundation in Boston had two shortwave stations, WRUL and WRUW. Also in Boston, Westinghouse had WBOS while shortwave broadcasts also emanated from Cincinnati through the Crosley station WLWO.

But the Allies were still out-shouted on the airwaves: there were by mid-1942 only fourteen shortwave transmitters (all privately owned and operated at that stage) in the United States and fifty in Britain; by contrast, Germany had begun the war with sixty-eight shortwave transmitters and by now, with the Italians and those of the occupied countries, had at its disposal more than a hundred.

The U.S. government took over shortwave station KGEI in San Francisco (which had been opened in 1939 by General Electric Corporation as a means of promoting GE products internationally). Under government control, KGEI carried broadcasts to Asia in English, Japanese, Mandarin, Cantonese, Tagalog, Dutch, Thai and Malay. The owners of San Francisco medium wave news station KSFO heeded the government appeal for help and, within three months, had a powerful shortwave station KWID up and running. KWID had a range of antennas allowing broadcasts to be directed to Alaska, the Far East, Australia and Latin America. In 1943 another transmitter was added and so was born KWIX. (Both stations were leased by the government until the end of the Korean war in 1953 and then dismantled.)

*Unidentified members of the staffs of KSFO and international shortwave station KWID, both operated by the Associated Broadcasters, are shown at the stations' shared transmitter site near Islais Creek in San Francisco in this undated photograph. (Staff from the Office of War Information, which advocated the creation of KWID, may also be present in this photograph.) The KWID shortwave antenna array was located in a field adjacent to the transmitter building. After it was dismantled, the KWID antenna site became a parking and storage lot for a tallow rendering plant.* John Schneider collection.

In Boston, station WRUL was taken over by the government in November 1942 and became Radio Boston to its shortwave listeners. Among the many services it was able to provide, it allowed young English evacuees to send messages to their families still in London; even before it became a government-run station, WRUL had broadcast messages on shortwave from the governments-in-exile of Belgium, Greece, the Netherlands, Norway and Yugoslavia to their

occupied home countries. Indeed, on 23 January 1942 it even broad-
cast in Luxembourgish featuring the son of Prime Minister Pierre
Dupong. The station broadcast in Tagalog to the Philippines. By
1943, the two transmitters were carrying programming compiled by
the Office of War Information in New York, in Albanian, German,
Arabic, Czech, Danish, Dutch, Finnish, French, Greek, Hungarian,
Italian, Norwegian, Persian, Polish, Portuguese, Spanish, Swedish,
Thai, Turkish and Serbo-Croat.

WRUL's two transmitters were located at Scituate, Massachusetts,
one with a radiating power of 50,000 watts, the other with 20,000
watts ( the second transmitter operating under the call-sign WRUW).
An example of the range of its signals is evidenced by the decision in mid-1942 to erect a new antenna array that would beam
WRUL broadcasts to Madagascar. It was reported by *The Wall Street
Journal* that the additional daily program in French would impress
upon the inhabitants of the island that if they persisted in supporting
Vichy they would suffer the consequences of Vichy's handing over
French territory to Japan, Italy and Germany. Once Free French
authorities took control of colonies, there was increasing broadcasting co-operation: Radio Brazzaville announced the times of French
broadcasts from WRUL, for example.

But what about the people listening?

The Nazis forbade listening to foreign broadcasts. The British
and Americans had no such fear of enemy propaganda, but hardly
encouraged people to tune in to those broadcasts. The Irish had not
even those qualms: *The Irish Times* published hour-by-hour listings of
news bulletins in English including those from Axis stations as well
as Allied. At 12.15 am Bremen radio was giving the German version
of the news in English. At 8.45 am Melbourne could be heard. At
11.00 am you had a choice between signals from Moscow, Rome
or Paris. At 2.30 pm Vatican Radio brought the latest church news
on shortwave, at 2. 50 pm the Japanese Radio Hsinking was audible,
and then at 3.30 pm the French radio station in Saigon. At 9.45 pm
there was English language news from Paris and Stockholm, and at
10.10 pm from Budapest.

According to *Time* magazine in July 1944 the 'weirdest listening post in all the world' was to have the radio turned on in Chungking. There was a twenty-four hour a day bombardment of propaganda that could be heard in the wartime Chinese capital. Much of the content was in English, clearly aimed at Americans working with Chinese forces and supplying them from Burma. Every afternoon at 4.00 pm there was a female presenter—in the then *Time*-style she was a 'brassy-voiced Japcaster'—who called herself Little Orphan Annie, interspersing American music with news bulletins highlighting bad news from the United States (such as floods, crimes and strikes). What struck George Grim, sent to Chungking by the State Department to help run local radio station XGOY, was the number of announcers who had almost faultless American accents. It certainly was a station format that would never work today: sessions other than spoken word programs consisted of classical music, including xylophone recitals.

## II

What more appropriate place, in China, for intrigue and shadowy figures than Shanghai? On 12 August 1945, as the Japanese were about to surrender, the Office of Strategic Studies (OSS) filed a report titled 'Shanghai - Counter-Espionage Summary'. Its forty pages conjure up a city riddled with intrigue, duplicity and shady Westerners. It catalogued the puppets in the city into three classes: first, the publishers, writers, news correspondents; second, bankers and men of the financial circles; and, third, gangsters and men of the underworld. It stated that Shanghai was the main centre for Japanese propaganda in 'Greater East Asia'. It was not just the Japanese. The Germans did not overlook this fertile field of propaganda and had at the time of Pearl Harbor, a radio station, magazine, daily newspaper and two news agencies in Shanghai, together with at least twenty-seven known agents engaged partially or wholly in propaganda.

In fact, it was a very confusing media scene. There were five shortwave stations operating from Shanghai. These included station XGRS which was controlled from the German legation to broadcast to the Pacific (it could be heard as far away as Australia). Its programs covered mainly the European war, but it transmitted in English, French, German, Russian and Chinese.

XGRS's English-language staff included John Holland, an Australian who had arrived in Shanghai in 1937, and Reginald Hollingsworth from Britain who entertained locals with a program called *Shanghai Walla Walla*, full of stories about the black market, speculation, racketeering and kidnapping. Hollingsworth was believed to have turned up in Shanghai from Japan after the war began. The station also employed Herbert Moy, a New York-born Chinese.

*The New York Times,* just hours before Japanese aircraft arrived on the Honolulu horizon, reported on another participant in the Shanghai broadcasting world and the cast of broadcasters working for the Japanese stations. 'Patrick Kelly is an announcer at the Shanghai broadcaster and as the sun rises over China he mocks the British, flashes Axis news of the Orient and Europe, advertises headache pills and cheers for Ireland'. The newspaper noted that Japanese short-wave stations were springing up all over China, 'from Canton to Peiping'. The Shanghai monthly magazine, *Twentieth Century*, was claimed by the OSS to be subsidised by German propaganda min-ister Joseph Goebbels. Then there was a German daily newspaper called *Noon Extra*. Station XIRS was run by the Italians, and again focused its coverage on the European theatre of war. Shanghai FF2 (or FEZ), controlled by Vichy interests, ran only news from France and broadcast no opinion.

Station XMHA was another interesting operation. This was American-owned and before December 1941 maintained a highly critical stance on Japan, even after Shanghai was controlled by Japanese troops. One broadcaster, Carroll Alcott, learned he was on the Japanese death list, after which he wore a bullet-proof vest, carried a gun and travelled in armoured car with guards. He quit Shanghai well before the Pearl Harbor attack. More pliant English-speakers were employed once Japan was at war with the Allies.

# 13.

# Picking up the pieces

WHEN THE WAR ENDED, Norway was faced with not only the problems of its own people but the fact that it had some half a million Germans, their prisoners of war and slave labourers, still on Norwegian territory. When in February 1946 the Oslo correspondent of *The Times* filed an update of the situation nine months after the Nazi surrender, there were still 40,000 people remaining in need of repatriation; most of those had their homes in what was by then the Soviet zones of Germany and Austria and clearly were in no hurry to return.

But the gravest problems for the Norwegians lay in the far north counties of Troms and Finnmark, the latter being the region that curves across the north of Sweden and Finland to meet the most northern section of Russia's eastern frontier. Together they cover an area equivalent to the combined size of Belgium and the Netherlands, and life was hard at the best of times with two months a year during which the sun did not shine. The Germans established an intensive network of military posts and airfields. The Germans had decided the region was too difficult to hold against the Soviet forces, and in their retreat had destroyed everything they could: all the

houses were burned to the ground and concrete foundations dyna-
mited; fishing boats—almost all 14,000 of them—were destroyed
and livestock slaughtered; the harbours and fjords were mined and
the lighthouses blown up. All 60,000 people living in the region
were forcibly evacuated. Every bridge was blown up, telegraph poles
felled, power lines pulled down, even fences were smashed. In the
town of Hammerfest, only one house remained standing. Twenty-
one hospitals were destroyed throughout the region. The Germans
even poured creosote on bread they could not carry with them.
Many people, when ordered by the Germans to leave, tried to save
their possessions by burying them, but very few of these were ever
found once the war was over. Children and old people tended to
succumb to disease; people had to constantly be on the move because
of frequent Nazi search parties.

The Times correspondent reported after travelling through
Finnmark that the destruction was terrible to see. Most farmers and
owners of small holdings traditionally kept cows, and there were also
large numbers of pigs and sheep. Almost all of these were killed by
the retreating Germans. Fortunately, some of the reindeer were able
to be driven to safety.

Further horror began when Soviet forces attacked in October
1944. The people of Kirkenes, where there was an iron ore mine,
hid in the underground shafts. But the Russians called a halt after
securing the most easterly region. The Germans ordered the remaining civilian population still under their control to board ships to be
transported south. When the war ended, they inhabitants returned
over June and July of 1945 and survived as best they could. As The
Times reported,

> The few churches (left standing) were immediately used to live
> in; elsewhere tents had to be used. They dug in the ashes of their
> homes and salvaged broken crockery and rusty cooking pots.
> They put up fences and gathered in the hay, but the possibility
> of remaining for the winter depended on the arrival of building
> materials and the barest necessities of life.

Ships could operate only in daylight due to the destruction of light-houses and navigation lights, and the land was littered with land mines. German military barracks in the south were broken into sections and sent north for quick reassembly as temporary housing (there being no building timber available in this far north region) and by the end of 1946 eight thousand such temporary buildings were in use.

With a month of the war in Europe still to go, *The Washington Post* in April 1945 reported under the headline 'Finnmark rises again despite Arctic night'. In the middle of the polar winter, with gales and snow a constant feature, the Norwegians had been getting on with the job of providing shelter to many thousands of people. The paper said that, when the Norwegian officials arrived from their London exile, they found the locals apathetic; they felt they had paid too high a price for liberation.

But Norwegian troops moved in, repairing roads and telephone lines. Within five weeks, several communities had their electric power restored. But, the paper said, the biggest boost was when a ship arrived carrying building materials, American trucks and supplies of fuel to run them, temporary wharves having been built.

A special department, the Finnmarkskontoret (for Finnmark Office), was set up in southern Troms. About 40,000 people were re-housed over the summer and autumn of 1945. Fishing boats had to be found to get the economy ticking over again.

Permanent reconstruction began in 1947. By June 1949 it was estimated by *The Scotsman* newspaper that the equivalent of £50 million would be needed to complete the work.

## II

Most of occupied Europe was in a state of devastation in 1945 (and beyond). But once the Germans had gone, at least the clean-up and rebuilding could begin. Or so the Luxembougers could have been forgiven for thinking.

The German assault through the Ardennes in 1945 (the Battle of the Bulge" as it is known) has been well documented, and most of what has been written leaves the impression that it was fought on just French soil. But Luxembourg was also embroiled in this action, and the German drive interrupted its reconstruction efforts. As *The Times* reported in February 1945, 'this second sally into Luxembourg by the Nazis, with the plundering especially of pigs, horses, and cattle and the devastation of towns and villages which accompanied it, has been a grievous blow to the small country'. Prime Minister Pierre Dupong estimated that the German occupation had, overall, cost Luxembourg about a third of its wealth, or more than seven billion Belgian francs.

Although its suffering has been dwarfed by the savagery that prevailed in German-occupied lands in the east, Luxembourg nevertheless went through a very bad time. The country itself was abolished by the Nazis and incorporated in the Reich; German was declared the official language and the publication of newspapers in any other language banned. The use of the words 'grand duchy' and 'state of Luxembourg' were similarly forbidden. Much of the pottery industry during the time of occupation was ordered by the Germans to provide plates and mugs for German households to replace those destroyed in Allied air raids.

When the Germans withdrew the second time they left little food behind them, and Luxembourg, for the first time, was no longer able to feed all its people. The returned government immediately set about repairing farm-workers' homes, getting its hands on tractors and implements, and plans were made for young people to spend two years working at farming before they could seek work in industry. The mining and metallurgical industries were in bad shape, and three-quarters of employees had no work to go to. In mid-February 1945, the Americans shipped 17,000 tons of high-grade metallurgical coal from the captured Anna mine near Alsdorf in Germany to the steel mills of Luxembourg, the first exports from occupied Germany and the beginning of what was termed 'reparations in kind'.

It was a full two months after the second German withdrawal before the U.S. Post Office said it could once accept items to be mailed to Luxembourg, although this was limited to letters under one ounce and non-illustrated postcards.

## III

By December 1947, Professor Johannes Fagginger Auer, a Dutch-born Unitarian minister who by then was Parkman Professor of Theology at the Harvard Divinity School, was reporting from his homeland for *The Christian Science Monitor*. Apart from all the war damage, the Netherlands had lost for the time being its biggest export market—Germany. 'It may be years before the German people will be able to afford the agricultural products which Holland exported in such vast quantities to the markets of Berlin, Cologne and Hamburg,' he wrote. In addition, the vast investments in the Netherlands Indies were shattered. A merchant fleet had to be created from scratch, but the shipbuilders were hard at work, and KLM was attempting to re-establish its air services.

In a later piece, discussing the rate of recovery in Holland, Auer noted that one of the greatest problems was the fact that trade with the East Indies had not rebounded. Holland was, he argued, essentially a trading country; it needed not only its own exports but those of the Netherlands Indies to earn the level of foreign exchange required for not only reconstruction but for consumer goods. He saw the East Indies' abundance in rubber, oil and many other commodities as the revenue salvation for the home country.

## IV

Inflation stalked parts of Europe as the war ended. At the 1945 annual meeting in London of the Ottoman Bank, Colonel E. Gore Browne described to shareholders how the Germans had systematically encouraged inflation in Greece. After the liberation, new drachma were issued at the equivalent of one to 50 billion of the old. At first, the new currency was fixed at 600 to the pound sterling, but by 1945 this had been adjusted to 2,000 to the pound.

In the week after the liberation, *Time* magazine on 13 November 1944 described the Greek financial system as laying prostrate. Finance minister Alexander Svolos tried to instil confidence in his country, citing the $175 million in gold reserves although the government could not get its hands on the yellow metal. No one was listening, apparently; *Time* reported that the value of a British gold sovereign rose from ten trillion drachmas to twenty-two trillion. Shopkeepers closed up, unwilling to sell for drachmas in return. Bartering was back. Bread sold for eleven billion drachmas a loaf.

Inflation was a fact of life as the war came to an end, and the money printing had been prodigious. Notes in circulation over the span of the war rose sharply: according to the League of Nations in 1945, note circulation in Greece was up more 312,422,000 per cent; even in non-belligerent Uruguay, money in circulation rose by twenty-five per cent. In Italy, Finland, Iceland, the Middle East, India and Japan the range was between 500 and 1,000 per cent. In the Baltics and Greece, paper money was losing its role as a means of payment. Croatia resorted to a barter system with farmers being paid for their produce in salt, cigarettes, tobacco and matches. In Syria, wholesale prices between July 1939 and December 1943 were up 800 per cent; in Turkey, a non-belligerent, prices rose close to 600 per cent by the end of 1943, although they started declining as 1944 wore on. Indian prices were up, in 1943, by 253 per cent on what they had been in 1939.

Greece's hyper-inflation was but a curtain-raiser for what was to come in Hungary. When Weimer Germany finally took action against its now infamous period of hyperinflation, the rate of conversion from old to new currency was at the rate of one trillion to one. In Hungary's case, in August 1946, the new notes were issued at the exchange rate of four hundred octillion to one. Whereas inflation had been a widespread problem after the 1914–18 war, only Hungary had a repeat episode well after the Second World War.

As with its Weimar precursor, the hyperinflation in Hungary began at what would later seem a modest pace. By June 1946, that pace became more frenetic. The *Manchester Guardian* in June 1946 described inflation as having reached 'fantastic heights'. In 1939 a loaf of bread had cost one-tenth of a pengo, the currency which had been introduced in 1927 to replace the then highly inflated korona; by June 1945, bread was costing one pengo. In April 1946, the equivalent amount of bread would have set you back ninety-six million pengo.

On the third day of January 1946, a correspondent of the *Christian Science Monitor* told his American readers that, because Hungarian notes were basically worthless, the only means by which to buy food was through barter. Parents in the city were sending their children to villages where food was available, the youngsters being provided with salt or boxes of matches to exchange for meals. By this time most city dwellers were existing on bread. Reporter R. H. Markham met a friend who he described as one of the most highly paid officials in Hungary. This fellow's family breakfasted on sugarless artificial tea, bread or potatoes, a tiny pat of butter and unsweetened jam. Lunch and supper consisted of soup and vegetables. The problem for government employees was that prices had increased roughly one hundred times faster than their salaries. The previous October, the Hungarian prime minister Bela Miklos received by way of monthly salary the amount of sixty-three thousands pengos; by January, that sum was insufficient to pay for a kilogram of apples.

Hungary, having joined the Axis powers, was forced in January 1945 to agree to pay reparations to the Soviet Union, Yugoslavia and Czechoslovakia to the tune of $300 million. But a more serious problem was that the taxation system broke down, and in the immediate post-war period only about ten per cent of government spending was covered by revenue collections. Borrowing had begun as soon as the pro-Nazi government fled Hungary taking the note printing plates and paper with them.

The stabilisation program ushered in the florint—the Hungarian rendering of the florin, which had been the currency of the Austro-Hungarian empire. You could collect a new florint at the bank by turning up with four hundred octillion pengos. Hungary was also able to recover twenty-two tonnes of gold that had been stolen by the Nazis; it was delivered to Budapest in Adolf Hilter's former private train. Tax rates were hiked—and made punitive to investors, with an eighty per cent rate on property income while companies were made to pay tax on revenues rather than profitability. By the end of 1946, ninety-six per cent of government spending was covered by government revenues.

In tandem with putting the brakes on inflation, the government went to great efforts to get goods into shops. William A. Bomberger and Gail E. Makinen in a 1980 article for an academic journal recount one transaction trail:

> Large quantities of Hungarian tobacco products, e.g., were sold in Vienna, where their price was high in dollars. The dollars were used to purchase cheap Polish sugar. The sugar was sold in Bucharest where its price was high in broken gold. The gold was taken to Budapest where it was struck by the Hungarian mint into gold Napoleons. These gold coins were then used to buy imported goods from Switzerland.

## V

Freed of the Japanese, the people of the colonies were lukewarm at best at the prospect of the return of their former colonial masters. On top of that, the metropolitan powers faced the enormous task of repairing colonial economies devastated by the occupation and battles.

Waldo Drake reported in February 1946 to *The Los Angeles Times* that the economy of the Netherlands Indies was virtually paralysed; communications between the hundreds of ports scattered around the archipelago were non-existent due to the destruction of the thousands of small coastal vessels that were the core of the Dutch merchant service. The development during the war of synthetic rubber raised the concern that the natural rubber industry would no longer be needed, a concern shared in British Malaya.

Before Pearl Harbor, as the newspaper recounted, this Dutch territory produced thirty-eight per cent of the world's rubber, twelve per cent of its tin, seventeen per cent of its tea, seventy per cent of its kapok, ninety per cent of its kapok and seventy-nine per cent of the world's pepper; in addition, it had been an important supplier of spices, resins, gums, fibres and palm nuts and oil.

Over on the Philippines, and after much bitter fighting, Japanese soldiers were cleared from the capital, Manila, on 3 March 1945, although it would be another five weeks before an army fort in Manila Bay was retaken. The Philippine economy was near collapse. Before the war, the country had sent more than eighty per cent of its exports to the United States, a trade worth $US256 million a year to the country. It would require a programme of nation rebuilding before the export business could be restored.

In the city's business district only two buildings remained undamaged, reported *Time* magazine in April 1945. The only supply of electricity in the central district came from army generators; the two hydro-electric plants owned by the Manila Electric Company, and located outside the capital, were still in Japanese hands. Shops were opening, but there were few goods available for sale although food was moving through the black market, eggs that officially cost three centavos were changing hands at fifty centavos. The U.S. military was distributing eight hundred thousand pounds of food a day. As the magazine pointed out, the absence of goods for sale led to a common attitude of reluctance to work, there being nothing to spend wages on. The Manila Stock Exchange was left as a name only.

For those living in the liberated areas of countryside, movement was almost impossible, all the buses having been destroyed. The jungle was closing in on sugar plantations that had closed during the Japanese occupation, and the narrow gauge railways that carried the sugar cane had often been dismantled by the Japanese troops as they retreated. The gold and copper mining industries were paralysed and there was concern that many underground shafts would have collapsed.

By mid-September 1945, an unnamed Australian reporter was filing reports picked up by *The Times* in London that in Singapore 'the city has returned remarkably quickly to nearly normal conditions'. Postal services were operating again, although they were described as 'skeleton' in nature, and trolley buses were moving through the streets. Free rice, sugar and salt were being issued by the newly returned British authorities. However, malaria was again rife, the disease having been stamped out before the war. And in the back streets, wrote the reporter, the shopkeepers 'with their myriad flies, are doing a thriving trade in food of unappetising appearance and in vegetables, bananas, pineapples, pawpaws, raw fish, dried whitebait, and various Chinese delicacies'.

On the Malayan peninsula, two resistance organisations, the Anti-Japanese Union and the People's Resistance Army, were organising labour gangs, distributing food, and policing areas until the regular forces returned. The historian and regional expert Paul H. Krastoska recalled that 'the Malayan population had not eaten well under Japanese rule, and people wanted to be liberated not only from Japan but also from tapioca'.

A committee set up under Sir Hubert Young, who had served successively as governor in Nyasaland, Northern Rhodesia and then Trinidad and Tobago before the war, had begun work in late 1943 on preparing estimates for goods and food that would be needed when the war ended in Burma, Malaya, British Borneo and Hong Kong. But these estimates, in the case of Malaya, both underestimated the population (partly due to the number of Chinese immigrants who had neither been born in the Malay states nor been naturalised, and therefore were not counted by officials) and overestimated the expected stockpiles of rice.

The situation was not made any better by the fact that many people who had more or less forcibly been required to grow veg-etables tore up their gardens after the Japanese surrender. And, once more, British targets for new production were unfulfilled: 200,000 acres of fallow rice land were to have been planted with off-season crops, but only 3,000 acres were actually sown, in addition to which the areas planted in tapioca and sweet potatoes declined significantly. Moreover, the owners of rubber and palm oil plantation estates were reluctant to get further involved in food production after having been forced to do so by the Japanese.

Black market operators filled some of the gap. But it was not until 1948 that rice production in Malaya regained its pre-war level.

## VI

To judge from the naval force that arrived in Hong Kong on 30 August 1945, the British government clearly wanted to make its presence felt and to assert that colonial authority was restored. The size of the fleet steaming into Victoria Harbour that day was not far off what constitutes the entire Royal Navy in the second decade of the twenty-first century. It included two carriers (*Indomitable* and *Venerable*), the cruisers *Swiftsure* and *Euralys* (along with the Canadian cruiser *Prince Robert*), six destroyers (*Kempenfelt, Ursa, Whirlwind, Quadrant, Tyrian* and *Tuscan*), seven submarines (*Selene, Supreme, Sidon, Spearhead, Solent, Seascout, Sleuth* and *Scotsman*), along with the submarine depot ship *HMS Maidstone*, seven Australian minesweepers and the hospital vessel *Oxfordshire*.

When British forces took possession again of the colony and its leased New Territories in that month, the population was drastically smaller than when they had surrendered to the Japanese. One report in *The Times* during 1947 estimated it at about half a million, well down on the figure of somewhere between 750,000 and one million for the level in 1937 although other estimates state it was anywhere between 400,000 and 600,000 compared to 1.6 million in 1941. Yet another puts a figure of 800,000 people facing the returning British. (By the time a census was conducted in 1961, the population had increased to 3,129,648.)

Within weeks of the British military government being put in place, the flood of people from the mainland began and was soon running at about one hundred thousand a month, the mainlanders fleeing an increasingly chaotic China riven by civil war between the Nationalist and Communist armies. The government in the colony had quickly set in place a system that established rice rationing, a stable Hong Kong dollar, and imposed the rule of law. *The Times'* correspondent reported the alarm at what was described as the unrestricted influx of Chinese in 1946 and 1947; however, many of these appeared to be former residents returning to the colony, these people having fled to the relative safety of the countryside across the border to escape the attention of the Japanese rulers in Hong Kong. The British were back, of course, due largely to a mainland power vacuum: both the Kuomintang and Communists were so preoccupied with fighting each other that they could not spare the energy and men to prevent the colonial administration being re-established.

Even with all the pressure on the recently liberated Crown colony, the British administration allowed the return of Indians who had taken refuge in the Portuguese colony of Macau. They had been under the protection of the British Consul, John Pownall Reeves, who kept the Union Jack flying throughout the war (the nearest other one was in Chungking at the British Embassy); he had been permitted by the Japanese to receive funds from London sent via Lisbon.

But, along with the returning former inhabitants, many other Chinese were moving into the British colony. Newspaper reports said monthly arrivals were running around 100,000 people, including many wealthy Chinese from Shanghai and southern China trying to preserve what they could of their assets and escape ruinous price inflation, blackmail and capricious tax levies.

But Hong Kong was initially unprepared to handle all these people. Much of the housing stock had been rendered uninhabitable by looting (and, to a lesser extent, by bombing). The water, electricity, gas, tram and bus services were in appalling shape. While the wealthy Chinese arriving in Hong Kong could pay exorbitant 'key money' to secure comfortable lodgings, most returning Britons had to make do with dormitories organised in hotels.

Even by 1947, many of the colony's manufacturing plants were still out of commission. These included sugar refineries, cement plants, textile and footwear factories; and those which had got back into production were inadequate in terms of providing jobs for all the newly arrived Chinese seeking work. Two years after the end of the Japanese occupation, the costs of rebuilding and supporting a fast growing population had seen the government deficit blow out to HK$115 million.

## V

A correspondent writing for *The Times* of London in August 1945 had no doubts about the gravity of the situation in British East Africa with all the returning native soldiers (or Askari, as they were commonly known) facing demobilisation. It was, he averred, the greatest social problem—and opportunity—for East Africa since the abolition of slavery. These men had been given a mental stimulus they had never previously known. They had seen many foreign lands and people, including the streets of London. They had been trained in new skills, from being medical orderlies to carpenters to lithographers and Education Corps instructors. The writer was concerned that thinking thus far had been circumscribed: how many medical orderlies could be absorbed into the health system, how many signalmen could be found jobs in the Post Office, for example. When the troops had been released from army service in 1918, the East African social structure had been changed little since the arrival of the European, so the men could go back to their villages and resume life as they had known it. Not so in 1945: the whole social structure had changed dramatically.

There was also concern in South Africa where 345,000 people, including 25,000 women, were facing demobilisation. Of that number, 122,000 were non-Europeans. The Union government had set aside 100 million South African pounds for demobilisation, and the decision was made that no service people would be discharged until a job had been found for them; all those employed before the war had to be reinstated for a minimum of one year. But about forty per cent had not been employed, and it was decided they would be kept on by the military at full pay until work could be found. Demobilisation committees were set up throughout the country, but much of the effort was made in the government sector. Some 3,300 jobs were created in South African Railways and Harbours, 1,700 in posts and telegraph, 1,000 in the police force, and another 3,000 throughout the public service. The Department of Lands was ordered to either settle on the land, or find jobs in agriculture and forestry, for another 3,000.

## VI

A correspondent of *The New York Times* had arrived in the first days of September 1945 at the Hotel Fujiya at Miyanoshita, within sight of Mount Fuji, an establishment which he described as having two swimming pools, games rooms and a fountain filled with gold-fish, and where white-coated waiters were available to guests. Much of the hotel was occupied by diplomats of Japan's Axis allies, each awaiting news of what was going to happen to them. They were provided with food, including white bread, and twenty Japanese ciga-rettes a day per person. There were 150 German, Thai and Italian diplomats. Former German ambassador Heinrich Stammer still did not know if he faced being arrested by the Allies and wanted to return to Germany to learn his situation, protesting to the journalist that he had never committed any war crimes. One of the Italian diplomats, Salvatore Merge, was accompanied by his New York-born wife Florence. Mrs Merge told the reporter that her greatest needs were for magazines, cigarettes, chocolates and news from New York. Sr Merge said the Japanese had treated the Italian diplomats well even after his country's surrender in 1943.

The next month saw *Time* chief political correspondent Manfred Gottfried wondering what was going to happen to Japan as a trading nation. 'Right now,' he filed, 'Japan is an industrial dust bowl'. The country pre-1937 needed five million tons of shipping to maintain peacetime trade; in 1945, only 450,000 tons were left afloat. Japan was out of oil, down to 5,000 tons of cotton for its textile sector and had only 40,000 bales of wool left.

Gottfried said, of all the problems, the hard one was what to do with the zaibatsu, the huge family-controlled conglomerates that controlled much of the Japanese economy and were driving forces in Japanese imperialism and the co-prosperity sphere. In the end, not all the much was to happen. Consider that the big four were Mitsui, Sumitomo, Yasuda and Mitsubishi. Only Yasuda was dis-solved, although one descendant of the family—Yoko Ono—would go on to fame and fortune in another realm. As Gottfried reported,

Mitsui had taken some severe blows: it had lost fifty per cent of its flour milling capacity, thirty per cent of its light metals capacity, forty per cent of its chemical manufacturing and most of its ships.

This must be qualified, however, that Japan still had industrial capacity beyond any of its Asian neighbours; and it still retained a large number of technically trained workers. A study in early 1946 published by *Far Eastern Survey* showed there were still some functioning steel rolling mills, blast furnaces and coke ovens. Japanese refineries in 1945 still had the capacity to produce 80,000 tons of copper ingots a year, along with refined tin, zinc, lead and copper. Aluminium capacity had suffered relatively little bomb damage. A report to the United States authorities showed Japan could still manufacture 1,000 railway locomotives a year and 17,000 freight wagons; in addition, 5,469 steam locomotives remained in working order. As many as 114 merchant ships over 5,000 gross tons were still afloat, and 794 smaller freighters. While the merchant fleet had suffered horrific losses, Japanese shipyards still intact could build 176 steel ships a year and sixty-seven dry docks remained in one piece. Telephones rang in 840,245 business premises or residences in Japan, still more than had been connected in 1931. The industrial devastation was not of the scale which Germany (and much of Europe) had witnessed.

The supreme commander General Douglas MacArthur ordered the break-up of the zaibatsu, but he did not reckon on the skill of the Japanese in obfuscating and evading regulation, nor the American commercial interests who were owed substantial amounts of money by the zaibatsu from pre-war trade dealings. You can see MacArthur's point: these huge enterprises were intertwined with government and political reform would also need the power of the top businessmen to be broken.

In 1945 the ten largest zaibatsu accounted for thirty-five per cent of Japan's paid-up capital, fifty-five per cent of the nation's bank assets, seventy-one per cent of loans and advances and sixty-seven per cent of trust bank deposits. At the end of the war, and notwithstanding the loss of so much of its capacity, Mitsui remained the world's largest private businesses with 338 companies and 1.8 million employees.

The Americans did have wins, Yasuda for one, with the dismissal of around 220,000 senior executives. But the real threats—which included Chinese demands for Japanese factories to replace its destroyed industrial base, and demands for Japanese reparations from Australia, Britain and the Soviet Union—were fobbed off by MacArthur. There would be no reparations, he decided. The Japanese industrial chiefs devoted their energies to ingratiating themselves with the new American masters, making friends; they were supported by United States banks and industries which had long relationships with the zaibatsu. The new finance minister, Keizo Shibusawa, was not exactly a clean-skin: he had been governor of the Bank of Japan in the latter years of the war and a substantial shareholder in the Teikoku Bank, formed by the mergers of the Mitsui and Dai-Ichi banks. He announced in October 1945 that the zaibatsu would 'spontaneously close shop'. Not even the Japanese press bought that line: the *Daily Asahi* labelled the announcement as merely 'reorganisation in disguise'. The procrastination continued until the whole zaibatsu issue was overwhelmed by the onset of the Cold War. The Americans now needed the Japanese as allies; of the 300 companies which had been slated for breaking up, only twenty suffered that fate.

One of the early decisions in the life of the occupation of Japan after the war was to close down all organs and businesses that had been part of Tokyo's plan to colonise Asia. Apart from taking control of a number of Japanese banks and other financial institutions, General Douglas MacArthur ordered that all institutions that had as their foremost purpose to finance colonisation and war production be closed. The Associated Press filed from Tokyo that these included the Manchurian Heavy Industry Development Company, branches in Japan of various banks including the Korean-based Bank of Chosen (which, the news report stated, had been the base for attempts to spread economic chaos in China before the Japanese invasion) and the Bank of Taiwan, which was described as having been the instrument of industrialists who exploited the Philippines, the Netherlands Indies and New Guinea. Mitsui and Mitsubishi were linked by this news agency report to that bank.

Meanwhile, the Japanese had to be fed. The country had its worst rice harvest in thirty years after typhoons and floods struck the defeated nation. By February 1946 the occupation authorities felt that they needed to intervene, telling the Japanese government to post an ordinance which allowed confiscation of rice crops in order to meet the need in the cities.

## VII

Meanwhile, even by December 1945, 9,000 Japanese troops were still on duty as railway guards near the port of Chinwangtao (now Qinhuangdao) working alongside the U.S. Marines. Japanese military and civilians still working for the railways were anxious to ingratiate themselves with the American forces, by then in command of the area. They saw the Americans, according to newspaper accounts, as protectors against the anger of the local Chinese who reportedly beat them and threw rocks. But the reporter found the Americans themselves to be angry, not just about having their former enemies as partners, but also about protecting British property in the form of the giant Kailan Mines Administration coal operations, located halfway between Tsientsin and Chinwangtao. The mining company argued that, if the U.S. Marines were to leave, the mine operations would have to close.

Nine months on, and the Marines were even less happy. They were reported to be disillusioned with the Kuomintang Government. Their sole task by this time was to guard the Chingwangtao to Tsientsin railway along with the associated (and 'filthy') settlements that housed the Chinese miners. These American troops had been living in bare shacks along the track but, in the expectation of staying another winter, were erecting barracks complete with electric lighting and bathrooms.

There were almost four million Japanese military and civilians in China at the time of Japan's surrender. Even after the surrender, Japanese generals retained control of the administration of Shanghai and Japanese soldiers continued to patrol the streets, treating the

Chinese with their customary arrogance. This went on for some months as the Japanese systematically looted the city, using trucks to haul away food and goods even after the arrival of the Nationalist Chinese forces. The sentiment pervading the Japanese forces in the country was that they had been defeated by the Americans, not the miserably led Kuomintang armies under Chiang Kai-shek, and they believed it would be dishonourable to surrender to Chinese forces they had continually defeated in the field.

Much of the Japanese army remained in China for almost a year after the end of the war, most of them retaining their weapons and still keeping order. The United States Embassy reported in January 1947 there were still about eighty thousand Japanese troops in Manchuria, all being supplied at the expense of the Chiang Kai-shek government, now back in Nanking. The Shansi provincial warlord, Yen His-shan, enlisted units of Japanese troops to fight the communists, and this they did until 1949. It was only one of several examples where Japanese soldiers went over to the Nationalist forces or worked with them. The motives were mixed: a fear of being unemployed or hungry back in Japan, fear of war crimes trials or a feeling of disgrace and not wanting to face their relatives and friends. There was also widespread belief among the Japanese that they could remain in China in some sort of partnership, both business and military. Japanese propagandists continued to foster the idea that Oriental people should resist any long-term influence by the Americans or Soviets. The Japanese harboured hopes of using China to restore Japan's economic and military power.

But it was not only military staying on: the Chinese recruited many Japanese technical civilian staff. They were offered special rations in Peking and in Mukden there was a government financed education programme provided to Japanese families. Overall, though, the majority of the civilians were not harnessed to China's economic needs and were eventually sent back to Japan. Moreover, while the Nationalist Chinese bigwigs and their cronies needed Japanese military help, they wanted to get their hands on the industries and businesses owned by their former conqueror.

## VIII

When the 1946 annual meeting of the Hongkong and Shanghai Banking Corporation was convened in the recently freed British Crown colony, the chairman Arthur Morse—after retailing the problems with re-establishing banking operations throughout what had been Japanese-occupied territories—then outlined the way in which he said Southeast Asia's clock had been set back after four years of 'Japanese depredations and the disorganisations of the war'. He said Burma's recovery was painfully slow, Java was still in turmoil, while uncertainties abounded regarding currency in the Philippines. However, he believed Malaya and Hong Kong were at an advantage in their currencies being linked to sterling.

Britain, after 1945, was able to re-establish many of its commercial interests throughout the empire, but it was another matter in Latin America. Between 1940 and 1950, the amount of British capital invested in that region fell by about half. According to the *South American Journal,* in 1939 British money in Latin American investments totalled £1,127 millions; by the end of 1949, those investments had shrivelled to £560 millions. The reduction was most dramatic in those countries which had the lion's share of British capital. In Argentina investments fell from £428.5 millions to £69.4 millions, followed in severity by Brazil, Chile and Mexico; the declines were slight by comparison in the rest of Central America, up a whisker in El Salvador and Paraguay, and only in Ecuador was it unchanged after ten years. The sectors most seriously affected were government bonds, railways and commercial banking.

Britain, after all, had in the nineteenth century taken on much of the task of financing the former Spanish and Portuguese colonies in South America. It was Barings and Rothschild banks that, between 1823 and 1825, provided the first wave of loans that were used to buy arms (to prevent Spanish reconquest) and also to finance mines in Bolivia and Colombia. Between 1815 and 1830 it was estimated that something in the order of £20 millions was subscribed in the London market for investment in Latin America. Another wave of investment

had begun in 1846 with the plan to construct the Central Argentine
Railway; by 1876, more than £27 millions had been invested by the
British in Argentina, accounting for ninety per cent of all foreign
money inflows into that country. By 1940, South America accounted
for about one-fifth of all British foreign investment.

But there were gains for British investors from the process of
Latin American governments wanting to take control of infrastructure and utilities owned by foreigners. In liquidating their investments
in various railway operations, the British got rid of some of the least
profitable of their assets. The only railway owned by British interests
that, in 1949, paid a dividend as much as six per cent was Dorada
Railway, which operated a short line in Colombia. At the end of the
war, British money was tied up in minority holdings of a number
of less satisfactory rail investments, including Mexican National
Railways, the International Railways of Central America and the
Guayaquil and Quito Railway.

Among the investments still retained by British investors after the
war were Montevideo Waterworks, Backus and Johnston's Brewery
in Peru, Rio de Janiero Flour Mills, Anglo-Ecuadorian Oilfields
and Pato Consolidated Gold Dredging of Colombia. Even those
companies that maintained their positions in Latin America felt the
impact of the conflict. Lobitos Oilfields, listed on the London Stock
Exchange, closed much of its capacity over three months of 1941
with its loss of markets; from March 1942 the Peruvian government
forbade shipments of oil to Britain. Of the company's five tanker
vessels, two were lost due to enemy action. Expatriate staff suffered,
too. Those stuck in Peru by the war suffered great strain without any
home leave, with some being sent to the United States for medical
treatment, *The Times* reported in 1945.

During the war, too, the British government made direct commercial agreements in areas which had previously been the preserve
of private commerce. Liebig's Extract of Meat Company reported to
shareholders in 1943 that Britain had negotiated directly with authorities in Buenos Aires for supply of meat at such a price that cattle
prices had risen to such a degree that it could run its canneries in
Argentina only at a loss. War restrictions had also made it a struggle to

obtain sufficient fuel and packaging materials with which to keep the factories operating. A severe winter in Argentina and Paraguay took its toll of calving and lambing rates, adding to the pressures of the war.

Of course, British investments suffered seriously, and not just in the war zones. In early 1953, *The Financial Times* was reporting that Portuguese legislators were casting their eyes over the British controlled utilities in their country, using the country's large sterling holdings accumulated during the war. Between 1938 and 1942, Portugal's imports had remained reasonably static, but exports had risen by 244 per cent, with sales to the Allied powers rising by 333 per cent. Those companies attracting particular attention from the legislators were Anglo-Portuguese Telephone (founded in 1887, the year of Queen Victoria's jubilee) and the Lisbon Electric Tramway, established in 1899.

An investigation in 1946 for the New York-based Foreign Policy Association found that, after the war, much of the German investment in Argentina was still intact. It noted that the regime headed by Colonel Juan Domingo Perón had waited until 26 January 1944 to cut diplomatic relations with both Germany and Japan. The authors of the report found that, after the break in diplomatic ties, German investment in Argentina remained high and had not been altered significantly since the war began in 1939. German money was in electric, metallurgical, steel and chemical industries along with sugar, cattle and other farming. German investment was estimated at two billion pesos (around $600 millions). However, with German's defeat only weeks away, the Argentines finally acted, confiscating about 150 subsidiaries of German companies and these were to be placed in the hands of Argentine owners. But it was scraping around the edges: the Americans argued that not one of the 108 major Axis enterprises (the ones suspected of working for German war interests) had been eliminated.

## IX

With the oceans, land masses and skies once again free of conflict, people wanted to be on the move. Businessmen needed to travel to expand their operations and do deals, visit those markets that had been closed to them for years; politicians and civil servants of the Allied victors had a lot of catching up to do with the liberated territories; civilians just wanted to shake off the strictures and confines of life in wartime.

Within a month of Japan's surrender, the Civil Aeronautics Board in Washington was drawing up plans for a radical restoration of Pan American Airways international operations. PanAm, or PAA as it was referred to in those days, should, the proposal outlined, extend its services from Midway Island in the Pacific to Tokyo, Shanghai, Hong Kong and Calcutta; it should also re-establish its flights to Manila, Singapore and Batavia. Officials wanted PAA to extend its service to New Caledonia on to Sydney. Meanwhile, Washington saw the need for Northwest Airlines to have aircraft depart from Chicago, stop en route in Canada, Alaska, the Aleutian islands, Paramushira in the Kuriles ( a planned thwarted when Soviet forces seized the island and its four airfields; Japan gave up sovereignty in 1951) and then complete their journeys to Tokyo and Shanghai.

X

American farmers, too, faced adjustment, although *The Wall Street Journal* in August 1945 proclaimed it would be 'with a gentler jolt' than to many other industries. For one thing, it was expected to mean a gradual end to labour shortages as men were released from the military and returned to their peacetime jobs. Secondly, factories which had been churning out aircraft and weapons would revert to making industrial machinery, farm equipment, and building materials, all by this time desperately needed by the farmers.

The consumer, too, would benefit. More and more items would go off the ration lists, the grocer's shelves would fill up again and restaurants would be able to expand their menus. Butter and cheese would become more plentiful in the shops. Meat was expected to be plentiful as farmers slaughtered animals to take advantage of the continuing high food prices.

The military would need to buy far less food: they could use the enormous stockpiles they built as insurance. The military had been accounting for sixteen per cent of all food produced and they were believed, the newspaper reported, to have a year's supply of some food items. Against this, of course, was the acknowledgment that Americans would have to help feed those in the destroyed nations. And the fact that some items would require replanting in those devastated lands; so there would be ongoing shortages of sugar, coconut oil and other oils.

The paper worried that the wool market could be disrupted. Much of the wool supplies were owned by the British government, but large parts of that were still sitting in storehouses in the southern hemisphere (Australia, New Zealand, Argentina) because there had not been the shipping capacity to move it. They hoped the British would not dump wool and shred the price.

# 14.
# Loose Ends

THE JANUARY 1941 BULLETIN from the League of Nations reported a general increase in the prices of imported products across Europe due to the British blockade and general shipping difficulties. Swiss imports for December 1940 had been fifty-four per cent greater in price than equivalent local products; Bulgarian import prices were twenty-two per cent higher and Spanish ones were twenty per cent up.

But here is a question: did anyone actually read this bulletin? Was anyone astonished that the League of Nations, established after being the fourteenth of President Woodrow Wilson's Fourteen Points to prevent another world war, was still toiling away? Germany and Russia had initially been banned from joining; America did not want to—it was official policy in Washington that any mail arriving at the State Department from the League of Nations was not to be opened, while the U.S. consul-general in Geneva was not allowed to make any public references to the body.

The league's decline had been triggered by a number of blows: Japan's invasion of China, the withdrawal of Germany and Italy and then the Second World War itself, the very event the prevention of which had been the league's *raison d'etre*. By 1941 it was irrelevant—but still in business.

Yet in 1942, for example, Britain's civil estimates still included £76,000 as that country's contribution toward expenses incurred by the league. The 1940 World's Fair in New York saw the League of Nations mount both its first and last pavilion at an international exhibition. Its building at the fair, tellingly, was placed on the very periphery of the international zone.

At the beginning of 1940, about three-quarters of the league's staff was dismissed. Those employees to go in the latest round included the sole Soviet official, an under-secretary-general, who was only two years into seven-year contract at $13,600 a year. On 10 January 1940, the *Chicago Tribune* ran a two-sentence item that had been filed the previous day from Geneva: 'The League of Nations will be twenty years old tomorrow. There will be no celebrations by secretariat members'. The *Christian Science Monitor* on that same day was considerably more generous with space: it reported praise for the league from the Finnish delegate for the backing that country had received over the Soviet invasion, and the paper reported that, of its total of sixty-two member countries, only ten had pulled out or been expelled. But that situation did not last long. Denmark declared the league a dead letter and said it would no longer pay its dues. By July 1940, Venezuela (which had joined in 1920) was gone, following in the footsteps over preceding months of Guatemala, Brazil, Costa Rica, Paraguay, Nicaragua, Honduras, El Salvador and Chile. Then, on 27 July 1940, the secretary-general, France's Joseph Avenol, resigned.

A report in late 1941 stated that the league continued to function in Geneva with an acting secretary-general and 'a reduced staff'. Much of the international body's activities, however, were being conducted outside continental Europe. The economic and financial section was based at Princeton University in the United States, while in Washington there had been opened offices of the league's Central Opium Board. The health committee was being run out of Geneva and Singapore, and the league's treasury was being managed from an office in London. The International Labour Office—one of the parts of the league to survive the war and become part of the United Nations (as did the Central Opium Board)—had established its headquarters in Montreal. Not everyone had given up on the league: in early 1941 South African prime minister General Smuts argued it had a future role to play. He was replying to Nationalist Party members in the House of Assembly who had criticised provision for a SA£25,000 vote for South Africa's contribution to the league. Smuts said the league would continue to provide a means for co-operation between nations.

In 1942 another report was issued, the World Economic Survey 1941-1942, which opened with the words: 'It is most fortunate, for both present and future students, that the Economic Department of the League has been able to continue so much of its work despite the war'.

As late as 1944, the league's health committee was the forum for agreement on an international standard unit of penicillin. The meeting in London was attended by representatives of Britain, the United States, Canada, Australia and Free France, along with medical specialists from South Africa and India. The discoverer of the drug, Sir Alexander Fleming, was adviser to the meeting. Also in that year, the Central Opium Board (with members from the United Kingdom, the Netherlands and China) met to discuss the control of narcotics as occupied countries were liberated.

The League of Nations was formally dissolved on 18 April 1946.


## Latvia's merchant marine

As the war in Europe began, Britain (and its empire) ruled the waves both militarily and in terms of merchant shipping, controlling twenty-one million tons of steam and motor vessels (of more than 100 tons displacement) out of a world total of 68.5 million. Germany and Japan owned 4.5 million tons and 5.6 million tons respectively. The Americans had 8.9 million tonnes of merchant marine. The losses were horrendous; of the twelve major merchant marine players, only Sweden (neutral) and the United States ended the war with a greater tonnage than in 1939. Norway had 4.8 million tons afloat, Holland 2.9 million tons. The Germans did not get it all: by the very nature of the business, many ships were operating in other parts of the world when their home countries were invaded. And the Allies bombed and sank what they could, which meant huge losses of shipping by France, Belgium, the Netherlands, Poland, Denmark, Norway, Greece, Yugoslavia, Romania and Bulgaria.

The three Baltic republics, however, were in a category of their own. As a result of the Hitler-Stalin pact that saw the two dictators divide Poland, the U.S.S.R was also given a free hand in the three republics to the north that had won their independence with the collapse of Tsarist Russia. The complication was that Washington did not recognise the Soviet takeover and the efforts of Moscow to retrieve the merchant ships flying the flags of Estonia, Latvia and Lithuania were thwarted. The U.S. District Court was confronted by a claim by the owner of the Estonian vessel *SS Kotkas* against its master, the latter refusing orders to return to Tallinn. The case was thrown out on the grounds that an executive order from the White House prevented transferring property of nationals of invaded countries. Several other attempts by the Soviet Union to get control of Estonian vessels at that time lying in American harbours were also frustrated.

Latvia was particularly hard hit by the Soviet invasion. Since 1918 enterprising Latvian entrepreneurs had built one of northern Europe's largest merchant fleets using existing and new vessels. Before the Second World War, Latvian ships carried some forty per cent of the country's seaborne trade which went through the ports of Riga, Liepaja and Ventspils. Latvia had sixteen steamship companies providing connections to northern European and British ports. They carried bacon, butter and wood products out, and brought back machinery, coal and consumer goods. By 1 January 1940, the Latvian merchant marine numbered 103 ships which was greater than the merchant fleets of Italy, Spain, Portugal or Poland.

In 2005, U.S. president George W. Bush made a speech in the Latvian capital, Riga. He recalled the story of the eight Latvian vessels that were in the Atlantic when the Soviet Union invaded the Baltic states. The entire Latvian fleet not in the home port was

ordered to return immediately. Sailors on those eight freighters—the
*Ciltvaira* and its sister ships *Everasma, Abagra, Everalda, Regent, Everelza,
Kegums* and *Everagra*—chose to remain at sea under the flag of free
Latvia. They recognised the Latvian embassy which remained open
in Washington even though the government and country it represented had ceased to exist. Two of the ships owned by the Latvian
Shipping Company happened to be at Rio de Janeiro when the order
came from Moscow. The *Ciltvaira* and *Everalda* were instructed, in
a messages from the Soviet embassy in Washington, to sail immediately for Murmansk. The ships were loading manganese for the
Ore Steamship Company of Baltimore and the United States Steel
Corporation respectively. They were held up for two months in
port as the Soviets tried through the Brazilian courts to get control of the ships. Eventually the ships' masters got clearance to sale
to their originally planned destinations. The eight ships were then
absorbed into the U.S. merchant marine after America entered the
war on 7 December 1941. As Bush recounted it in the presence of
the presidents of Estonia, Latvia and Lithuania, by the end of the war,
six of the Latvian ships had been sunk (only the *Kegums* and *Everagra*
survived) and more than half the sailors lost. One American town
renamed a street Ciltvaira to honour one of those sunken ships.

In 2003, the managing editor of a small newspaper in Nags Head,
North Carolina, became interested in the *Ciltvaira* story after reading
reports in the Riga newspaper, *The Baltic Times*. As Sandy Semans
then of the *Outer Banks Sentinel* told it, on 19 January 1942, U-123
torpedoed the Latvian vessel off that North Caroline town. The
crew had farewelled their loved ones two and a half years earlier.
After refusing to return to Latvia on Soviet orders, the men knew
they faced a death sentence if ever they fell into Soviet hands. (Some
other ships did obey the order from Moscow. The crew on the *Hercog
Jacob* was one such: it sailed from Peru to Vladivostok where it was
renamed *Soviet Latvia* and for fifteen years transported prisoners to
the Gulag labour camps on Sakhalin Island).

The *Ciltvaira,* of 3,779 tons and built in 1904, was heading from
Norfolk to Savannah with a load of paper. At 5.00 am, a torpedo hit

the boilers, killed two stokers (a third had gone on deck for a breath of air) and flooded parts of *Ciltvaira*. Two hours later a passenger line refused to stop, obviously fearing the U-boat was still in the area. The Latvian tramp steamer, although badly damaged, was still afloat when, at 9.00 am a Brazilian freighter arrived and tried to tow the helpless vessel but the towrope broke (some of the *Ciltvaira* crew having by this time rowed back to the ship and reboarded her). Then a Standard Oil tanker, *Socony Vacuum*, arrived on the scene and took off the thirty surviving crewmen and the ship's pet cat and dog. *Ciltvaira*, her back broken, sank.

*Ciltvaira after being torpedoed.* U.S. Navy.

## A sad end for the last of the great chartered companies

The chartered company was a popular commercial tool employed by European powers in their colonising periods. It enabled private capital to speed the process of acquiring control of areas and to bear the costs of so doing. In return, the companies were given mandates to create and profit from trade. The empire in its early stages rode on this mercantile back. The Hudson's Bay Company, the East India Company, the Royal Niger Company, the German New Guinea

Company and the Dutch East India Company were among the many such legendary businesses.

Six weeks after Pearl Harbor, a reporter and a photographer from Britain's *Picture Post* magazine were allowed into the boardroom at the last surviving such enterprise, the British North Borneo (Chartered) Company. The company had been incorporated by royal charter in 1881 and been given, 'in perpetuity', control of what became known as British North Borneo, all 77,700 square kilometres of it (and now the Malaysian state of Sabah). The decision by a British order-in-council followed the land being granted to the company three years previously by the sultans of Brunei and Sulu. Company rule—the last surviving such charter in the world—was to resume briefly after Japan's defeat but on 26 June 1946 the territory became a British crown colony.

By the time of the Japanese invasion of Borneo and much else in Southeast Asia, the board still represented a British ruling elite that seemed, then, able to endure for centuries. The president, 72-year old Major-General Sir Neill Malcolm was a product of Eton and Sandhurst, had served with the Argyll and Sutherland Highlanders and seen duty on the North-West Frontier and in Uganda, South Africa and Somaliland. His younger brother was vice-president: the sixty-four year old Sir Dougal Malcolm was described as a well-known figure in the City and was also a director of the British South Africa Company. Also on the board was Robert Buxton DSO, whose directorships included Anglo-Egyptian Oilfields Ltd; Buxton had served in the Sudan Civil Service and in the Great War had been a lieutenant-colonel with the Imperial Camel Corp, a unit raised in 1916 to re-assert British control against insurgents in the Libyan Desert. Another director was the Hon. Mountstuart William Elphinstone, aged seventy, who had been at the War Office in the previous conflict with Germany.

The company was not in the usual business of trading; rather, it derived its revenues from customs and excise duties, court fees, stamp duties, royalties and land rents, along with telegraph revenue, money from wireless stations, wharfs, lighthouses and from a 200 kilometre railway line. Law and order was maintained by a police force of 550 (many Indian) with five European officers (the 1931 census showing the company territory contained 270,000 people of whom just 340 were Europeans). Total revenue in 1938, expressed in Straits Settlements dollars, was $3.25 million; customs duties and excise taxes accounting for $1.78 million of that; land revenue generated $341,156. The railway line, unlike many around the world even then, seems to have been profitable, with revenues of $224,876 against expenditures of $149,329.

British North Borneo was a valuable source to Britain and its dominions of rubber, along with timber and tobacco. The Japanese invasion had taken its toll on all of this, and of the company: the *Picture Post* report quotes Sir Dougal Malcolm telling the board that 'since our debentures are so illiquid, I think we should concentrate on things as they'. The demand for rubber due to the war had grown substantially, and this produced greater excise earnings and higher revenue on the company's railway until the Japanese invasion. Two months before Pearl Harbor, *The Financial Times* declared of the company, which had just declared a profit of £169,000, that 'so far the prospects for the current year appear to be somewhat more promising than those of a year ago'.

By the time of the annual meeting in London on 30 June 1942, Sir Neill told shareholders of the measures the company had put into effect when it became clear that Japan was going to war. The local population was encouraged to grow as much food as possible and large stores of rice were laid in. 'I fear that they (the stores) have gone to fill other mouths than those for which they were intended,' he told the meeting. On receipt of the news that the Japanese landing was taking place, the order had gone out to destroy all vessels and other items which might assist the invader. The company was reduced thereafter to living from its other investments. At the 1944 annual meeting, Sir Neill disclosed that the offices had been leased out to the Overseas Farmers Co-operative Federation for £1,250 a year. Meanwhile £20,000 was being earned from investments although the company had been receiving requests for assistance from the families of officers who had stayed at their posts and been captured. The president also made a commitment to phase out the opium trade in North Borneo by 1950 even though it would cost £40,000 a year in lost revenues.

Not much was left standing in North Borneo by the time the Japanese were dislodged: Allied bombing took a heavy toll. Buildings, equipment, boats—all were destroyed. The chartered company, as a result, decided upon selling out to the Crown. After prolonged negotiations, made more complex by the absence of any audited accounts since the end of 1940, there was a deadlock; eventually the matter went to arbitration, the amount of £1.4 million being paid for all the assets of the company. At least the shareholders redeemed something, which is more than the majority of North Borneo inhabitants managed to achieve. Apart from the devastation, rubber prices had fallen substantially and rice was short; even by mid-1946, the daily rice ration was down to two ounces per person.

## World War II—Great Business for Postage Stamps

The regular postage stamps column in *The New York Times* of 7 September 1941 noted that at least 1,245 new issues had by that stage of the war resulted from the conflict. Of that number, about five hundred new stamps were occupation postal paper issued in Alsace, Belgium, Bohemia and Moravia, Bulgaria, Danzig, and other territories seized by the Nazis. The article was illustrated with several new stamps, included one issued by the General Government in Poland. Latvia and Lithuania, now under Soviet occupation, had also seen new stamps, as had Vichy France and the then twenty-two possessions loyal to that government. However, about 120 new varieties had appeared from the Free French areas including Cameroon, French Equatorial Africa, French India, French Oceania and New Caledonia.

The dominant image on stamps issued by the Vichy government was that of Marshall Philippe Petain; otherwise, Vichy stamps tended initially to show non-controversial images, such a farmer harvesting wheat. Vichy also issued stamps for parts of the French colonial empire, although few were actually sent to those places. Most of them were sold to collectors in France; they were designed primarily to persuade the French that the colonial possessions were at one with Vichy. Indeed, the Vichy government continued to issue stamps for French Equatorial Africa, French Oceania, French India, Reunion, Madagascar and the Cameroons even after those territories had switched their loyalties to De Gaulle and the Free French movement, which in turn issued their own stamps for the colonies on their side, so that by 1944 these places were represented by two differing sets of stamps, one printed in London, the other in France.

Meanwhile, once German troops had crossed into and subdued Belgium, the Belgian Congo suddenly could not replenish its supplies of postage stamps. So the authorities in Leopoldville turned to South Africa to print their stamps, with ten new issues in ten denominations being produced in Pretoria (all carrying a picture of the King Albert Monument which stood in Leopoldville). Belgian East Africa (now split into Rwanda and Burundi) also issued its own stamps.

The Danes, who lived under the least oppressive Nazi occupation, seemed free to keep issuing postage stamps reflecting their own identity. In 1941, for example, an issue came out honouring the Danish navigator Vitus Bering, after whose discovery the Bering Strait was named.

The war also saw the Portuguese embrace stamp collecting. As *The Economist* reported in 1942, stamps for purchase were displayed in a wide variety of retail outlets, especially stationers, tobacconists, lottery dealers and in bazaars. But the Portuguese were mainly interested in issues from the German Reich, including occupied Poland, and from Italy (including its issues for Somaliland and Eritrea). They particularly liked, it seems, stamps bearing the images of Hitler or Mussolini. The magazine's correspondent noted that none of the stamps, which were often sold in sheets, bore postmarks, indicating they were printed for collectors and the writer suspected that this was a useful way for the two Axis powers to raise some much needed foreign currency.

## The mail doesn't always get through

After the occupation and partition of France between the German sector and Vichy, postal services were often the only link between the two zones; they were also the way contact could be kept with those in prisoner-of-war camps and with the outside world. From October 1940 inter-zonal correspondence could be carried out only by use of *cartes familiales* for personal messages and *cartes commerciales* for business matters; these had pre-printed messages on the reverse with spaces for particular details of the sender, so little information could be transmitted by their use. A card of the time contains the stricture:

> '*After having completed this card specifically for correspondence with members of the family follow the instructions—Never write outside the lines. Attention: All cards not for family members will not be sent and will probably be destroyed*'.

It then had spaces for the place of despatch, the date and then you could select whether you wanted to report that the news about a person was they were either in good health, tired, ill or wounded, a prisoner, dead—or whether there was no news of them; then you could indicate whether you needed provisions or money, where luggage should be sent, and details of work or school and what mail you had received. As a final salutation, you As a final salutation, you could underline either 'Kind thoughts' or 'Kisses'.

This restriction was lifted in 1941, but postal items were routinely checked by the Germans. Eventually, French people were able to write letters to recipients in Germany, Italy, Belgium, Holland, Denmark, Norway and the Balkan countries.

Even with all the stamps, quite often the mail did not get through due to the constant merchant shipping losses. *The Times* ran regular reports to alert its British readers that letters and parcels they had sent would not reach their destinations due to enemy action. On 25 April 1941, for example, the lost outward mails included:

- Letters and printed papers for Barbados, Grenada, St. Lucia, St. Vincent, Trinidad and Tobago and Venezuela (Cuidad Bolivar only) posted between 8 January and 15 January;

- Letters and printed papers for Aden, Kenya and Uganda, Mauritius, Northern Rhodesia, Nyasaland, Seychelles, South Africa, Southern Rhodesia, Tanganyika Territory and Zanzibar, posted between 4 March and 6 March.
- Letters for Belgian Congo and China (Kunming only), French Somali Coast, Portuguese East Africa and Saudi Arabia, also posted between 4 March and 6 March.
- Magazine post for Canada and Newfoundland mailed between 10 March and 14 March.

Notwithstanding the chaos and ever-moving front lines, and the deaths of postal carriers in no man's land, the Chinese Post Office continued to ensure that the mail gets through between what is known as 'Occupied China' and 'Free China'. Shanghai was firmly in Japanese hands; Chungking was the capital of the Nationalist Government. But letters moved between the two cities even though taking twenty-one days to do so. Camels and wheelbarrows are among the conveyances used to carry the post overland when it was not possible to use motor lorries.

## The war is over—well, nearly

Ending all the states of war was somewhat more difficult. The United Kingdom and Thailand remained at war until an agreement was signed in Singapore on 1 January 1946. The Siamese authorities renounced claims made over parts of northern Malaya; these areas had been ceded to Britain in 1909 by what was then known as Siam. In early September 1945, the Supreme Commander, Southeast Asia, Lord Louis Mountbatten, issued a proclamation terminating Thailand's territorial gains and returning the four states to British Malaya. The authorities in Bangkok also agreed they would not build a canal linking the Indian Ocean with the Gulf of Siam without British approval (a proposal still being talked about today but as yet unimplemented). London also exacted a demand that Siam deliver rice surplus to that country's needs up to a maximum 1.5 million tons. Under pressure from occupying Japanese forces, Siam had on 25 January 1942 declared war on Britain and the United States.

## And what to do with all that capacity?

The war ended, but production did not. The Americans were faced with an industrial and agricultural sector working at full capacity and large parts of the world market lying prostrate.

China was one possibility as a market for industries switching to peacetime production. In 1940, the United States-China trade was worth about $95 millions; silver, silk, tung oil, bristles, rugs and skins came eastwards to American ports, while ships took to China cotton, leaf tobacco, automobiles, tractors, coal, iron and steel, kerosene, machinery, paints and varnishes. China, in a report to the United Nations Relief and Rehabilitation Administration said that, over the following twenty-four months, it would need $2.5 billion in goods to help rebuild the shattered country. The Nationalist government was clearly living in a dream world, announcing plans to building 40,000 miles of new railways as well as a highway system, new inland waterways and develop a domestic airline system. Had this occurred, then it would have soaked up a good deal of what was now surplus industrial capacity in the United States.

Meanwhile, one of the top priorities as seen from Washington was to get new equipment to the tin mines in Malaya and the Netherlands Indies. Mines in those countries supplied about seventy per cent of global needs and American planners were concerned about shortages hitting their own industries. The United States was also eager to get oil production restarted in Borneo.

# Postscripts

In a final swoop to cover loose ends, here is a chapter-by-chapter follow-up which closes off various incidents and developments related in this book. Consider it a 'what-ever-happened-to?' resolution.

### Introduction: The 'Strength through Joy' resort.

This was the resort described in the introduction which was intended for German workers located off the Baltic Coast of the Reich on the island of Rügen, and was near completion when Hitler ordered the invasion of Poland.

After the war, the resort was plundered by the Red Army and then made part of a restricted zone for the East German army's use. As for the cruise ships, both were sunk in 1945—the *Gustloff* with an estimated 10,582 refugees from the eastern areas by a Soviet submarine and the *Ley* by Royal Air Force bombers. As for the tourism resort at Rügen, in July 2011 the German daily *Die Welt* visited the site, still Germany's largest single building, all six storeys high and four kilometres long. (The original plan had been for a highway built along the succession of roofs.) As reporter Dankwart Guratzsch noted, it can be photographed in its entirety only from the air. The report was on the occasion of the German youth hostel movement opening a 400-bed operation in the old Nazi complex. Even with all those 400 beds, the hostel occupied only one-third of one block, Block Five. The newspaper report noted that the new youth hostel had spelled out in the first edition of its brochure that the facility was on the site of the Nazi leisure agency's dream; not surprisingly, it was decided to omit that fact from subsequent editions.

## Chapter 3: The Race for Resources

The Trepca lead and zinc mine which had been of so much interest to the Germans was nationalised by the Yugoslav government after war, and eventually the company was partly compensated. Trepca Mines Ltd was liquidated in 1957 after the final—but unsatisfactory to the company—payment was received from Belgrade. But the shareholders did at least get something. By contrast, in the case of the Bor copper operation, this had had been sold by its French owners during the Vichy regime to the Germans. In 1945, Yugoslavia announced it regarded Bor as expropriated German property and nationalised it.

## Chapter 4: Putting Food on the Table

On 16 October 1945, Minister of Food Sir Ben Smith announced Britain would resume importing wine from Europe and the Dominions. For the remainder of the year, the country would import 1,400 tons of wine each from South Africa and Australia and, for 1946, each of the Dominions would ship seven thousand tons of wine and six hundred tons of brandy. The British West Indies would ship seven thousand tons of matured rum for the remainder of 1945 and about five thousand tons in 1946. Cyprus was to send a total of 1,150 tons of wine through to December 1946.

Smith said that advanced negotiations were under way for the resumption of wine and spirit imports from France, and the first consignment would consist of the cheaper Burgundy and Bordeaux varieties of wine, champagne and brandy. Five hundred tons of wine would be imported from Algeria before the end of 1945 and 'large quantities' in 1946, while Spain, Portugal and Madeira would between them send three thousand tons of sherry, one thousand tons of port and one hundred tons of Madeira wines.

Maximum prices were also to be set: twelve shillings and six-pence for Australian port and sherry types, with brandy selling for between £1 10s and £1 16s. West Indies rum could be sold at up to twenty-seven shillings a bottle and French vintage champagne at £1 15s. Algerian wine was clearly going to be attractive to British drinkers, with a maximum price set of eight shillings.

Booze had been one of the early casualties of war. In September 1939 the Alcohol Monopoly Board in Finland, which had been created to control the industry upon the 1932 abolition of prohibition, upped the already high price of alcohol by another fifty per cent, thus making it affordable only to the wealthier Finns. The locals complained that this was just profiteering by the board as most of the liquor sold in Finland was produced domestically. The country was, even this early into the war, already experiencing shortages of sugar, salt, gasoline, cotton, coffee and tea.

But it was to get worse. In Finland, the price of vodka tripled between 1938 and 1944. In Norway, alcohol rose in price every year, and from 1940 until after the end of German occupation there was also rationing. Drinking rose in Sweden as purchasing power rose during the war, and wages in Denmark rose faster than the price of vodka.

## Chapter 5: The Advantages of Empire

It took a long time for the Dominions after 1945 to wake up to the fact that all their sacrifices would soon be forgotten by Britain. The book issued in 1945 by the Ministry of Information, *What Britain Has Done, 1939-1945* (referred to in the introduction to this work), should have been seen as the first clue. It was not called *What Britain and its Empire Have Done*. Then there was the subsequent Labour government's abandonment of 'east of Suez', and then the European Common Market. Visit the Imperial War Museum in London and you have to search hard for signs that every part of the empire pulled its weight, both in terms of fighting forces and material aid. "Empire? What Empire?" seems to be the museum's attitude.

* * *

Being a British colonial subject was often no protection even of itself. Aden, that small but critically positioned staging post for steamers plying the India route, was home to a small Jewish community— many were Yemeni Jews, and few from Baghdad and Persia, and even a handful of German Jews. When the Italian residents left at the start of the war, the Jews took over the piece goods trade. But the events in Palestine after the war spread to Aden. So great did Arab hostility become that the British surrounded the Jewish quarter with barbed wire. In December 1947, Arab youths set fire to the Jewish girls' school and Jewish-owned cars. In all, eighty-two Jews were killed, the majority of Jewish shops looted. By 1949, most of the Jews had emigrated to Israel.

## Chapter 6: Italy, the Weakest Link

In 1953 Associated Press filed a report from Tripoli on the plight of the Italian settlers still remaining in now independent Libya. Many of the 45,000 remaining had been born in Libya, most being concentrated in Tripolitania; Italian farmers were scattered throughout the province and Italians made up a large proportion of Tripoli's urban population. There would be no jobs for their children when they finished school: the Libyans needed all available jobs for their own. Yet Italy was experiencing a very high rate of unemployment, so there were few prospects for anyone contemplating moving back to the mother country.

## Chapter 8: The slow strangulation of Germany

When Germany invaded the Netherlands, the colonial authorities in the Netherlands Indies acted immediately to seize German cargo vessels. So what happened to those ships? The *Nordmark* became the *Mandalika* and was sunk on 19 March 1941 by U-105 while sailing from Surabaya to Belfast with a cargo of sugar. *Rendsburg* was renamed *Toendjoek* and was scuttled at Tadnjung Priok, although it was raised in 1943 by the Japanese and renamed *Tango Maru*, only to be sunk in the Java Sea by the submarine *USS Rasher*. *Vogtland* took the name *Berakit* and was sunk by Japanese submarine I-37 while en route from Colombo to Durban, while *Casell* became *Mendenaw* and was sunk in 1942 by U-752 while carrying stores from New York to the British in Egypt (it was going via Cape Town and the Suez Canal). The *Essen*, which became the *Terkolei*, was sunk by U-631 while crossing the Atlantic from New York with a cargo of zinc, wheat and mail while *Naumburg* was renamed *Kentar* and then subsequently was torpedoed and sunk by U-155 while carrying manganese ore from India to Saint John, New Brunswick, Canada.

* * *

The British occupation of the Faroe Islands, intended to prevent Germany using the Danish territory as a base, constituted one of those near-forgotten episodes of the war.

The main force of British troops to occupy the Faroes were Scottish regiments and, as *The Scotsman* initially decided to portray it, things could not have been more cosy. The tone was set by recounting the British takeover: it involved an unarmed officer with a small group landing from a motor launch and calling on the Danish governor to explain what was about to happen. Readers were told that the Faroese took well the news that there would be censorship and all contact with Denmark was being severed. The British brought all their food (apart from fish, which was sourced locally) in by ship and there was the added bonus that the local people were also given access to far more meat than that to which they had become accustomed. British troops were reportedly warmly welcomed into homes, and the locals even shared their last supplies of schnapps, the reporter claimed. Moreover, the Faroese were said to have developed a liking for the bagpipes. And there were frequent football matches between local and forces teams.

By 1944, however, the newspaper was giving a more realistic picture. Dr Gavin Henderson of the University of Glasgow had been to the islands to deliver a series of lectures intended, it was reported, to help alleviate the boredom being experienced by troops stationed there. He found considerable concern about the number of local women who had married British servicemen; by itself, 128 such marriages did not seem that many, but in a population of 28,000 people—the capital, Tórshavn, was home to about 3,000—it had considerable social impact, the Faroese worried how these girls would be replaced. Henderson also commented on the overall impact: the Faroese until 1940 had lived a life more redolent of the Victorian era than the twentieth century, and he saw the islanders suffering from the same adjustment problems that had afflicted the people of the Orkneys and Shetlands. Until the British arrived, few of the islanders had ever seen a film; the young women, though, quickly developed a liking for products such as silk stockings and lipstick. But it was no great shakes for the troops, either. Henderson said their enemy was not the Germans but the weather—the continual gales, the rain, the long winter nights, the isolation and the boredom.

* * *

The *Chicago Daily News*, which had reported plans by Berlin to established a postal union through its conquered territories, suffered the fate of most afternoon newspapers. It closed in 1968.

## Chapter 9: Co-prosperity—or bust

The policy of colonisation between 1935 and 1940 had seen 310,000 Japanese emigrate; many had moved to Korea and Formosa in the previous decades. But more went during the war as farmers were lured to Manchukuo and Japanese companies took over business operations in the Netherlands Indies, Thailand and other places.

When the war ended in August 1945 there were some seven million Japanese stranded throughout Asia, including 750,000 in Korea.

## Chapter 12: Winning—then losing—the propaganda war

When the war began, all shortwave broadcasting in the United States was conducted by private companies. Along with the creation of the Voice of America, Washington in 1942 took control of the private stations. The first to return to private ownership was WRUL, being handed back on 18 December 1945, its first non-government shortwave broadcast featuring Gertrude Lawrence (then playing in Boston). There seems to have been considerable reluctance on the government's part to relinquish control of the stations.

# Bibliography

## Introduction

DiNardo, R.L., and Bay, Austin, "Horse-Drawn Transport in the German Army", *Journal of Contemporary History*, Vol. 23 No. 1, Jan. 1988

Edgerton, David, *Britain's War Machine. Weapons, Resources and Experts in the Second World War*, Allen Lane, 2011.

Hastings, Max, *All Hell Let Loose*, HarperPress 2011

League of Nations, *World Economic Survey*, Geneva 1945

Ministry of Information, London, *What Britain Has Done 1939–1945*, 1945

Nakamoto, Hiroko as told to Pace, Mildred Mastin, *My Japan 1930–1951*, McGraw-Hill Book Co, New York, 1970

Olorunfemi, A., "Effects of War-Time Trade Controls on Nigerian Cocoa Traders and Producers, 1939–45: A Case-Study of the Hazards of a Dependent Economy", *The International Journal of African Historical Studies*, Vol. 13, No. 4 (1980)]

Royal Institute of International Affairs, "Raw Materials and Foodstuffs: the United Kingdom's Strategic Position", *Bulletin of International News*, Vol. 16 No. 17, 26 August 1939

Spode, Hasso, "Fordism, Mass Tourism and the Third Reich: the 'Strength Through Joy' seaside resort as an index fossil", *Journal of Social History*, Vol. 38 No. 1 Autumn 2004

Strauss, Frederick, "The Food Problem in the German War Economy", *The Quarterly Journal of Economics*, Vol. 55 No. 3, May 1941

*Washington Post*, 20 March 1942

## Chapter 1: Last Glimpses of Peace

Duranti, Marco, "Utopia, Nostalgia, and World War at the 1939–40 New York World's Fair", *Journal of Contemporary History*, Vol. 41 No. 4 Oct. 2006

Gropman, Alan L, *Mobilising U.S. Industry in World War II*, Diane Publishing 1996

Herman, Arthur, *Freedom's Forge: How American business produced victory in World War II,* Random House 2012

Napier, Russell, *Anatomy of the Bear: Lessons from Wall Street's Four Great Bottoms*, Harriman House 2007

*Auckland Star*, 4 September 1939
*Chicago Tribune*, 6 September 1939, 24 April 1940
*Christian Science Monitor*, 15 August 1940
*The Economist,* 10 May 1941
*The Irish Times*, 3 September 1939
*Los Angeles Times*, 25 August 1939, 2 October 1939, 3 September 1939
*The New York Times*, 3 September 1939, 28 January 1940, 4 June 1940
*Picture Post*, 26 August 1939
*Sydney Morning Herald*, 28 August 1939
*The Times*, 16 January 1940

## Chapter 2: Japan's Wasted Years

Bai Gao, "The State and the Associational Order of the Economy: The Institutionalisation of Cartels and Trade Associations in 1931–45 Japan", *Sociological Forum*, Vol. 16 No. 3 2001

Biggs, Barton, *Wealth, War and Wisdom,* John Wiley & Sons, 2010

Cohen, Jerome B, "The Japanese War Economy 1940–1945", *Far Eastern Survey*, Vol. 15 No. 24, 4 December 1946

Department of the Interior, *Fishery Leaflet 157*, December 1945

Gage, Eugenia, "Industrial Development in Formosa", *Economic Geography*, Vol. 26 No. 3. July 1950s

Gurda, John, "Profits and Patriotism: Milwaukee industry in World War II", *The Wisconsin Magazine of History*, Vol. 78 No. 1, 1994.

Han, Yu-shan, "Formosa Under Three Rules", *Pacific Historical Review*, Vol. 19 No. 4, November 1950

Herman, Arthur, *Freedom's Forge: How American Business Produced Victory in World War II*, Random House 2013

Kratoska, Paul H., "Labor Mobilisation in Japan and the Japanese Empire", in *Asian Labor in the Wartime Japanese Empire*, Kratoska (Ed), M. E. Sharpe, New York, 2005

Marx, Daniel Jr, "Shipping Crisis in the Pacific", *Far Eastern Survey*, 5 May 1941

Miller, Edward S, *Bankrupting the Enemy: The U.S. Financial Siege of Japan Before Pearl Harbour*, Naval Institute Press 2007

Phillips, R.T., 'The Japanese Occupation of Hainan', *Modern Asian Studies*, Vol. 14 No. 1, 1980

Rice, Richard, "Economic Mobilization in Wartime Japan: Business, Bureaucracy and Military in Conflict", *The Journal of Asian Studies*, Vol. 38 No. 4, Aug 1979

Shillony, Ben-Ami, "Universities and Students in Wartime Japan", *The Journal of Asian Studies*, Vol. 45, No. 4, August 1986

Shizume, Masato, *Japanese Economy during the Interwar Period: Instability in the Financial System and the Impact of the World Depression,* Institute for Monetary and Economic Studies, Bank of Japan, 2009

Singleton, John (Ed), *Learning in Likely Places: Varieties of Apprenticeships in Japan*, Cambridge University Press, 1998.

Wolborsky, Stephen L., "Choke Hold: The Attack on Japanese Oil in World War II", thesis presented to Air University, Maxwell Air Force Base, Alabama, 1994

Yasuba, Yasukichi, "Did Japan Ever Suffer from a Shortage of Natural Resources Before World War II?", *The Journal of Economic History*, Vol. 56, No. 1, September 1996

Yeh, Wen-hsin, *Wartime Shanghai*, Taylor & Francis, London 1998

*The Financial Times*, 14 June 1938
*The New York Times*, 2 June 1941
*Time*, 15 December 1941, 24 April 1939
*The Times*, 12 February 1941, 1 April 1939

## Chapter 3: The Race for Resources

Barnes, Kathleen, "British Antarctic Whaling Upsets Japanese Hopes", *Far Eastern Survey*, Vol. 9 No. 4, 14 February 1940

"B.S.K.", "Raw Materials: Germany's Potential Sources of Supply", *Bulletin of International News*, Vol. 16 No. 16, 12 August 1939

Dumett, Raymond, "Africa's strategic minerals during the Second World War", *Journal of African History*, 1985

Ginsburgs, George, "The Soviet Union, the Neutrals and International Law in World War II". *The International and Comparative Law Quarterly*, Vol. 11, No. 1, January 1962

Hayward, Joel, Hitler's Quest for Oil: the Impact of Economic Considerations on Military Strategy, 1941–42, *The Journal of Strategic Studies*, Vol. 18 No 4

Holdaway, B.N., "A Specialist Wartime Industry: the Development of the Linen Flax Industry in Marlborough", Journal of the Nelson and Marlborough Historical Societies, Vol. 2, No. 1, 1987

Kroener, Bernhard R, Muller, Rolf-Dieter and Umbreit, Hans, *Germany and the Second World War: Volume V/II: Organization and Mobilization of the German Sphere of Power—Wartime Administration, Economy and Manpower Resources 1942–1944/5*, Oxford University Press, 2003

Leitz, Christian, *Economic Relations Between Nazi Germany and Franco's Spain 1936–1945*, Oxford University Press 1996

Morales, Jonathan, *Oil and World War II on the Eastern Front*, Academi.edu

Oyebade, Adebayo, "Feeding America's War Machine: the United States and economic expansion in West Africa during World War II", *African Economic History*, No. 26, 1998

Roberts, A.D., "Gold Boom of the 1930s in Eastern Africa", *African Affairs*, Vol. 85 No. 341, October 1986

Stokes, Raymond J, "The Oil Industry in Nazi Germany", *Business History Review*, Vol. 59 No. 2, Summer 1985

Wendt, Paul, "The Control of Rubber in World War II", *The Southern Economic Journal*, Vol. 13, No. 3, January 1947

Tønnessen, Joh N. and Johnsen, Arne Odd, *The History of Modern Whaling*, University of California Press, 1982.


BBC News, 9 August 2010

*Chicago Tribune,* 22 April 1942

*Christian Science Monitor,* 2 April 1942, 6 March 1943

*The Economist,* 6 September 1941, 18 October 1948

*The International and Comparative Law Quarterly,* Vol. 11, No. 1

*International Herald Tribune,* 14 November 2006

*Los Angeles Times,* 9 October 1939, 31 October 1943

*The Manchester Guardian,* 16 December 1942

*The New York Times,* 2 July 1940

*Straits Times,* 2 February 1935

*Time,* 21 February 1944

*The Times,* 13 January 1940

*Washington Post,* 28 October 1939, 13 November 1939,

*The Wall Street Journal,* 25 August 1937, 15 May 1940, 15 July 1944

## Chapter 4: Putting Food on the Table

Bloch, Kurt, "Silk production curbed", *Far Eastern Survey*, Vol. 10 No. 10, June 1942

Brassley, Paul; Segers, Yves; Van Molle, Leen, *War Agriculture and Food*, Taylor and Francis 2012

"B.S.K.", "Raw Materials: Germany's Potential Sources of Supply", *Bulletin of International News*, Vol. 16, No. 16, 12 August 1939

Collingham, Lizzie, *The Taste of War: World War II and the Battle for Food*, Allen Lane 2011

Danquah, Francis, "Japan's Food Farming Policies in Wartime Southeast Asia: the Philippine example, 1942–1944", *Agricultural History*, Vol. 64 No. 3, 1990

Dung, Bui Minh, "Japan's role in the Vietnamese Starvation of 1944–45", *Modern Asian Studies*, Vol. 29 No. 3, July 1995

Giltner, Philip, "Trade in Phoney Wartime: The Danish–German 'Maltese' Agreement of 9 October 1939", *The International History Review*, xix, 2: May 1997

Johnson, Chalmers A., *MITI and the Japanese Miracle*, Stanford University Press, 1982.

Johnston, Bruce F., *Japanese Food Management in World War II*, Stanford University Press 1953

Kratoska, Paul H, "Post-1945 Food Shortage in British Malaya", *Journal of Southeast Asian Studies*, Vol. XIX, No. 1

Lund, Joachim, "Denmark in the European New Order", *Contemporary European History*, Vol. 13 No. 3, August 2004.

Sanders, Paul, *German Black Market Operation in Occupied France–Belgium 1940–1944*, Ph.D dissertation, University of Cambridge, 1999

Taylor, Lynne, "The Black Market in Occupied Northern France 1940–41", *Contemporary European History*. Vol. 6, No. 2. July 1997

Tooze, Adam, *The Wages of Destruction: The Making and Breaking of the Nazi Economy*, Allen Lane 2006

Yeong, Jia Chia, *Grow More Food Campaign*; paper compiled by National Library Board of Singapore.

*Chicago Tribune,* 16 February 1942, 12 May 1940
*Glasgow Herald*, 14 June 1939
*Christian Science Monitor,* 21 August 1942
*Milwaukee Journal*, 8 April 1944
*New York Times*, 4 June 1942, 25 May 1945
*The Scotsman,* 29 October 1941

*The Times*, 15 April 1940, 4 June 1942
*Washington Post,* 11 August 1940

## Chapter 5: The Advantages of Empire

Bromby, Robin, *German Raiders of the South Seas: The extraordinary true story of naval deception, daring and disguise 1914–1917*, Highgate Publishing 2012. (Originally published in 1985 by Doubleday Australia)

Crocker, Chester A, "Military Dependence: the Colonial Legacy in Africa", *Journal of Modern African Studies*, Vol. 12 No. 2. June 1974

Fetter, Bruce, "Changing War Aims: Central Africa's Role 1940–41 as Seen from Leopoldville", *African Affairs*, Vol. 87 No. 348, 1988

Gadesen, Fay, "Wartime Propaganda in Kenya: The Kenya Information Office 1939–1945", *The International Journal of African Historical Studies*, Vol. 19 No. 3, 1986

Grajdanzey, Andrew J, "India's Economic Position in 1944", *Pacific Affairs*, Vol. 17 No. 4, December 1944

Killingray, David, "The Idea of a British Imperial African Army", *The Journal of African History*, Vol. 20 No. 3 1979

Laband, John, *Daily Lives of Civilians in Wartime Africa: from slavery days to the Rwandan Genocide*, Greenwood Publishing Group, 2009.

Little, Marilyn, "Colonial Policy and Subsistence in Tanganyika 1925–1945", *The Geographical Review*, Vol. 81 No. 4, October 1991

Moresco, Emanuel, *Colonial Questions and Peace,* Columbia University Press, 1939.

Olukojo, Ayodeji, "Buy British, Sell Foreign: External Trade Control Policies in Nigeria during World War II and its Aftermath 1939–1950", *International Journal of African Historical Studies*, Vol. 35 No. 2/3

Samasuwo, Nhamo, "Food Production and War Supplies: Rhodesia's beef industry During the Second World War 1939–1945", *Journal of Southern African Studies*, Vol. 29 No. 2, June 2003

Thomas, Martin, "Deferring to Vichy in the Western Hemisphere: the St Pierre and Miquelon Affair of 1941", *The International History Review*, Vol. 19 No. 4, November 1997.

*Chicago Tribune,* 14 March 1941
*Christian Science Monitor*, 27 August 1941, 13 July 1942
*The Economist,* 17 February 1940, 22 February 1941, 20 September 1941
*The Indian Express,* 5 May 1941
*Los Angeles Times,* 26 December 1941
*The Manchester Guardian,* 10 September 1940,

*The New York Times,* 25 August 1941, 12 October 1941
*Picture Post,* 19 February 1944
Reuters, 6 December 1940
*The Scotsman,* 22 January 1940, 26 August 1940
*The Spectator,* 3 January 1941
*Time,* 18 May 1942
*Times of India,* 4 January 1941, 18 December 1940, 7 December 1941
*Washington Post,* 16 January 1949

## Chapter 6: The Axis Loses in Latin America

Burden, W.A.M, *The Struggle for Airways in Latin America,* Ayer Publishing 1977

Jones, Clarence, "South America and the War", *Proceedings of the American Academy of Arts and Sciences,* Vol. 75, No. 1

Loveman, Brian, and Davies, Thomas M., *The Politics of Antipolitics: The Military in Latin America,* Rowan & Littlefield, Boston, 1997

McCann, Frank D., *Brazil and World War II: The Forgotten Ally,* published by Estudios Interdisciplinarios de America Latina y el Caribe, University of Tel Aviv. McCann was emeritus professor of history at the University of New Hampshire.

Pineo, Ronn F., *Ecuador and the United States: Useful Strangers,* University of Georgia Press, 2007

Rippy, J. Fred, "German Investments in Argentina", *The Journal of Business of the University of Chicago,* Vol. 21 No. 1

Sommi, Luis Victor, *Los Capitales Alemanes en Argentina,* Claridad, 1943


*Chicago Tribune,* 22 September 1942
*The Christian Science Monitor,* 31 January 1942
*The Economist,* 10 May 1941
*Far Eastern Survey,* Vol. 6, No. 6, 17 March 1937
*New York Times,* 6 September 1938, 1 October 1939, 11 August 1941, 4 April 1942
*Time,* 19 September 1938

## Chapter 7: Italy—the weakest link

Bosworth, R.J.B., *Mussolini's Italy: Life under the Dictatorship,* Allen Lane, 2005

Ebner, Michael R., *Ordinary Violence in Mussolini's Italy,* Cambridge University Press, 2010

Fowler, Gary L, "Italian Colonisation of Tripolitania", *Annals of the Association of American Geographers,* Vol. 62 No. 4 1972

Helstosky, Carol, *Garlic and Oil: Food and Politics in Italy*, Berg 2004

Morgan, Philip, *The Fall of Mussolini: Italy, Italians and the Second World War*, Oxford University Press 2007

O'Hara, Vincent P, Dickson, W. David, Worth, Richard, *On Seas Contested: The Seven Great Navies of the Second World War*, Naval Institute Press, 2014

Paulirella, Eugenia, *Fashion Under Fascism: Beyond the Black Shirt*, Berg, 2004

Sadkovich, James J, "Understanding Defeat: Reappraising Italy's Role in World War II", *Journal of Contemporary History*, Vol. 24 No. 1

Schreiber, Gerhard, Stegeman, Bernd & Vogel, Detlef, *The Mediterranean, South-east Europe and North Africa 1939–1941; From Italy's Declaration of Non-belligerence to the entry of the United States into the War*, Oxford University Press 1995

Simons, Geoff., *Libya: The Struggle for Survival*, Palgrave Macmillan 1993

Tate, R. H., "The Italian Colonial Empire", *Journal of the Royal African Society*, Vol. 40, No. 159, April 1941

Townley,. Edward, *Mussolini and Italy*, Heinemann 2002

Van Creveld, Martin L, *Hitler's Strategy 1940–1941*, CUP Archive, 1973


*Christian Science Monitor,* 25 December 1937, 15 November 1938, 6 March 1941

*The Economist,* 18 October 1941, 30 May 1942

*The Financial Times,* 9 August 1939, 22 December 1942

*Manchester Guardian*, 10 February 1942

*The New York Times*, 17 December 1939

*Time*, 3 November 1941

*The Times*, 20 January 1940

## Chapter 8: The Slow Economic Strangulation of Germany

Briggs, Herbert W., "Non-Recognition in the Courts: The Ships of the Baltic Republics", *The American Journal of International Law*, Vol. 37 No. 4, October 1943.

Caruana, Leonard, and Rockoff, Hugh, "An elephant in the garden: The Allies, Spain and oil in World War II", *European Review of Economic History*, Vol. 11, No. 2, August 2007

DiNardo, R.L. and Bay, Austin, "Horse-Drawn Transport in the German Army", *Journal of Contemporary History*, Vol. 23, No. 1, 1988

Harrison, Mark, "The Volume of Soviet Munitions Output, 1937–1945: A Re-evaluation", *The Journal of Economic History*, Vol. 50, No. 3, September 1990

Joint Intelligence Sub-Committee (Britain), *Some Weaknesses in German Strategy and Organisation, 1933–1945*

Klemann, Hein A. M., "Dutch Industrial Companies and the German Occupation 1940–1945", *Vierteljahrschrift für Sozial-und Wirtschaftsgeschichte*, 93 Bd, H.1, 2006

Leitz, Christian, *Economic Relations Between Nazi Germany and Franco's Spain 1936–1945*, Oxford University Press 1996

Ministry of Information, *What Britain Has Done 1939–1945*, London, 1945

Royal Institute of International Affairs, "Outline of Military Operations", *Bulletin of International News*, Vol. 19, No. 3

Sanders, Paul, "Economic Draining—German Black Market Operations in France 1940–1943", Academia.edu

Schwendemann, Heinrich, "German–Soviet Economic Relations at the time of the Hitler–Stalin Pact 1939–1941", *Cahiers du Monde russe*, Vol. 36, No. 1–2

Tierney, Dominic, *FDR and the Spanish Civil War: Neutrality and Commitment to the Struggle that Divided America*, Duke University Press, 2007

Tooze, Adam, *The Wages of Destruction: The Making and Breaking of the Nazi Economy*, Allen Lane 2006

Vajda, Ference A, *German Aircraft Industry and Production 1933–1945*, McFarland, 1998

Weiss, Kenneth G, *The Azores in Diplomacy and Strategy, 1940–1945*, Centre for Naval Analyses, Alexandra, Virginia, 1980

Willoughby, Malcolm Francis, *The U.S. Coastguard in World War II*, Ayer Publishing 1980

Wylie, Neville, "British Smuggling Operations from Switzerland 1940–1944", *The Historical Journal*, 48, 4, 2005


*Chicago Daily News,* 23 October 1942

*Christian Science Monitor,* 15 August 1940

*Daily Telegraph,* 27 March 1940

*The Economist,* 10 May 1941

*Los Angeles Times*

*The New York Times,* 15 May 1940, 1 October 1942

*Picture Post,* 18 May 1940

*The Scotsman,* 26 August 1940, 22 January 1944

*Time,* 31 May 1943

*The Times,* 30 March 1940, 6 January 1942, 6 April 1943, 6 August 1943

*Washington Post,* 8 June 1943

## Chapter 9: The Co-Prosperity Illusion

Australian War Memorial, *Fighting in Timor, 1942*.

Chen, Ta-yuen, "Japan and the Birth of Takao's Fisheries in Nanyo 1895–1945", Working paper No. 139, Asia Research Centre, Murdoch University, Perth. Nov. 2006.

Cornelius-Takahama, Vernon, Singapore Infopedia, National Library of Singapore (on the Cathay cinema in Singapore)

Emerson, Rupert, "The Dutch East Indies Adrift", *Foreign Affairs*, Vol. 18, No.4. July 1940

Fisher, Charles A., "The Expansion of Japan: A study in Oriental geopolitics", *The Geographical Journal*, Vol. 115 No. 4/6, Apr–Jun 1950

Gage, Eugenia, "Industrial Development in Formosa", *Economic Geography*, Vol. 26, No. 3

Goto, Kenichi, "Japan and Portuguese Timor in the 1930s and early 1940s", published by Waseda University

Grajdanzoy, A. J, "Japan's Co-prosperity Sphere". *Pacific Affairs*, Vol. 16, No. 3.

Kratoska, Paul H., *The Japanese Occupation of Malaya: a social and economic history*, C. Hurst & Co, 1998.

Kratoska, Paul H., "Banana Money: Consequences of the Demonetisation of Wartime Japanese Currency in Malaya", *Journal of Southeast Asian Studies*, Vol. 23, No. 2, September 1992

Lee, Robert, "Crisis in a Backwater: 1941 in Portuguese Timor", *Lusotopie: enjeux contemporains dans les espaces lusophones: Lusophonies asiatiques, Asiatiques en lusophonies*, Volume publié avec le concours du CNRT (Paris) et de la Comissão nacional para as Comemorações dos Descobrimentos portugueses (Lisbonne), 2000

Omar, Maritsa, Article on Singapore amusement parks, Singapore Infopedia.

Phillips, R.T., "The Japanese Occupation of Taiwan", *Modern Asian Studies*, Vol. 14, No. 1, 1980

Smith, Norman Dennis, *Resisting Manchukuo: Chinese women writers and the Japanese population*, UBC Press, 2007

Swan, William L, "Japan's Intentions for its Greater East Asia Co-Prosperity Sphere as Indicated in its Policy Plans for Thailand", *Journal of Southeast Asian Studies*, 1, March 1996

Tan, Bonny, Article on Goodwood Park hotel, Singapore Infopedia

Tarling, Nicholas, Britain, Portugal and East Timor in 1941, *Journal of Southeast Asian Studies*, Vol. 27, No. 1, March 1996

Yasuyuki, Hikita, "Japanese companies' inroads into Indonesia under Japanese military domination", in *Bijdragen tot de Taal-, Land- en Volkenkunde* (or

*Journal of the Humanities and Social Sciences of Southeast Asia and Oceania).* Vol. 152, No. 4. 1996

*The Economist,* 23 September 1939, 29 March 1941, 20 June 1942, 22 September 1945
*Foreign Affairs,* Vol. 18, No.4. July 1940
*Los Angeles Times,* 24 September 1940
*The New York Times,* 7 March 1937, 16 October 1941
Singapore Infopedia, National Library of Singapore

## Chapter 10: Japan Feels the Squeeze

Allen, Matthew, "Undermining the Occupation: Women Coalminers in 1940s Japan", University of Wollongong, Research Online
Bernd, Martin, *Agriculture and food supply in Japan during the Second World War,* Sonderduck aus der Albert-Ludwigs-Universität Freiburg
Garon, Sheldon, "Luxury is the enemy: Mobilising savings and popularising thrift in Wartime Japan", *Journal of Japanese Studies,* Vol. 26, No.1, Winter 2000.
Griffiths, Owen, "Need, Greed and Protest in Japan's Black Market 1938–1949", *Journal of Social History,* Vol. 35, No. 4. Summer 2002
Havens, Thomas R. H, "Women and War in Japan, 1937–45", *The American Historical Review,* University of Chicago Press, Vol.80, No.4, October 1975
Nakamoto, Hiroko as told to Pace, Mildred Mastin, *My Japan 1930–1951,* McGraw-Hill Book Co, New York, 1970
Rossiter, Fred J, "Japan's New Rice Problem", *Far Eastern Survey,* Vol. 11, No. 12, June 1942
Sato, Shigeru, "Indonesia 1939–1942: Prelude to the Japanese Occupation", *Journal of Southeast Asian Studies,* Vol. 37, No. 2, June 2006

*Christian Science Monitor,* 14 November 1944
*The Economist,* 29 March 1941
*New York Times,* 19 March 1945
*Picture Post,* 27 September 1941
*Washington Post,* 20 March 1942

## Chapter 11: China—Japan's critical failure

Chin, Rockwood Q.P, "The Chinese Cotton Industry under Wartime Inflation", *Pacific Affairs,* Vol. 16, No. 1. March 1943.
Coble, Parks M., *Chinese Capitalists in Japan's New Order: The Occupied Lower Yangzi 1937–1945,* University of California Press 2003

Craw, Sir Henry, "The Burma Road", *The Geographical Journal*, Vol. 99 No. 5/6, May–June 1942

Fitzgerald, Patrick, "The Yunnan–Burma Road", *The Geographical Journal*, Vol. XCV, No. 3, March 1940

Mitter, Rana, *China's War With Japan, 1937–1945: The Struggle for Survival*, Allen Lane 2013

Morton, W. Scott and Lewis, Charlton M., *China: its history and culture*, McGraw-Hill Professional, 2005.

Tamagna, Frank M, "The Financial Position of China and Japan", *The American Economic Review*, Vol. 36 No. 2, May 1946

Young, Arthur N., *China's Wartime Finance and Inflation 1937–1945*, Harvard University Press 1965

Zhou, Yuan and Elliker, Calvin, "From the People of the United States: The Books for China Program during World War II". *Libraries & Culture*, Vol. 32, No. 2. Spring 1997.

*Christian Science Monitor*, 16 October 1943, 1 November 1943, 12 April 1944

*Los Angeles Times*, 24 May 1940

*New York Times,* 3 September 1939, 13 September 1944

*Singapore Free Press and Mercantile Advertiser*, 18 May 1939

*Time*, 13 November 1939, 9 November 1942

## Chapter 12: Losing the Propaganda war, too

Arsenian, Seth, "Wartime Propaganda in the Middle East", *Middle East Journal*, Vol. 2 No. 4, October 1948

Central Intelligence Agency, declassified documents

Gadesen, Fay, "Wartime Propaganda in Kenya: The Kenya Information Office", 1939–1945, *The International Journal of African Historical Studies*, Vol. 19, No. 3, 1986

Head, Sydney W, "British Colonial Broadcasting Policies: The Case of the Gold Coast", *African Studies Review*, Vol. 22, No. 2, Sept. 1979

MacDonald, Callum A, "Radio Bari: Italian Wireless Propaganda in the Middle East and British Countermeasures 1934–38", *Middle Eastern Studies*, Vol. 13, No. 2. May 1977

Mendelssohn, Peter D., *Japan's Political Warfare*, Taylor & Francis, Oxford 2010

Menefee, Seldo C., "Japan's Psychological War", *Social Forces*, Vol. 21, No. 4. May 1943

Murray, Jacqui, *Watching the Sun Rise; Australian reporting of Japan 1931 to the fall of Singapore* Jacqui Murray. Lexington Books, Washington DC, 2004

Ogilvie, John W. G, "The Potentialities of Inter-American Radio", *Public Opinion Quarterly*, Vol. 9, No. 1. Spring 1945

Associated Press, 18 September 1945
*The Irish Times*, 19 January 1940
*New York Times*, 4 December 1938, 7 December 1941
*Sunday Dispatch*, 21 February 1943
*Time*, 30 January 1939, 1 June 1942, 24 July 1944
*Wall Street Journal*, 18 May 1942
*Washington Post*, 20 May 1942, 4 August 1943,

**Chapter 13: Picking up the Pieces**

Bennett, Martin T, "Japan's Capacity to Produce", *Far Eastern Survey*, Vol. 15, No. 9, May 1946
Bomberger, W.A. and Makinen, G.A, "Indexation, Inflationary Finance and Hyperinflation: The 1945–1946 Hungarian Experience", *Journal of Political Economy*, Vol. 88, No. 3, 1980
Bomberger, William A. and Makinen, Gail E, "Hungarian hyperinflation and Stabilisation of 1945–1946", *The Journal of Political Economy*, Vol. 91, No.5. October 1983.
Hunter, Janet (ed) *Japanese Economic History 1930–1960*, Routledge, 2000
Kratoska, Paul H, "The Post-1945 Food Shortage in British Malaya", *Journal of Southeast Asian Studies*, Vol. XIX, No.1, March 1988
Louis, Wm. Roger, "Hong Kong: The critical phase", *The American Historical Review*, Vol. 102, No. 4. October 1997
Lund, Diderich, H, "The Revival of Northern Norway", *The Geographical Journal,* Royal Geographical Society, Vol. 109, No.4/6. April/June 1947
McCann, Frank D., *Brazil and World War II: The Forgotten Ally*. Published online by Tel Aviv University
Vaughn, T.D. and Dwyer, D.J, "Some Aspects of post-war population growth in Hong Kong", *Economic Geography*, Vol. 42, No. 1. 1966
Website of the Museum of Reconstruction for Finnmark and North-Troms
Olson, Martin, *The Silent Trust: Life Story of Dr Sandor Mihaly*, AuthorHouse, 2011

*Christian Science Monitor*, 1 October 1945, 18 December 1947
*Daily Asahi*, 22 October 1945
*Foreign Policy Reports*, 1 February 1946, Foreign Policy Association
*Manchester Guardian,* 8 April 1946
*Time* magazine, 13 November 1944

*The Times,* 21 August 1945

*Los Angeles Times,* 22 February 1946

*The New York Times,* 5 September 1945, 27 December 1945

*Time,* 15 October 1945

*The Times,* 21 August 1945, 20 September 1945, 25 October 1945, 1 February
    1946, 5 August 1946, 7 March 1947

*The Wall Street Journal,* 11 August 1945

*The Washington Post,* 1 April 1945


**Chapter 14: Loose Ends**

Hoisington, William A. Jr, "Politics and Postage Stamps: The Postal Issues of
    the French State and Empire 1940–1944", *French Historical Studies* Vol. 7
    No. 3, Spring 1972

Kahin, George McT, "The State of North Borneo 1881–1946", *The Far
    Eastern Quarterly,* Vol. 7 No. 1. Nov 1947.


*Chicago Tribune,* 10 January 1940

*Christian Science Monitor,* 10 January 1940, 3 December 1943, 17 December
    1945

*Die Welt,* 5 July 2011

*The Economist,* 20 June 1942, 15 September 1945

*The Financial Times,* 13 October 1941

*Los Angeles Times,* 3 March 1953

*The Times,* 17 October 1945

# Index

Abyssinia   83, 120, 121, 123
Aden   266
African Pioneer Corp   90
Agriculture   74–75, 91, 92
   Denmark   64–69
   France   71
   Italy   121
   Japan   75–76
   United States   248–249
Albania   124, 126
Alcohol   264–265
Algeria   87, 91
Angola   92
Antarctica   59–60
Argentina   108, 119
   German investments in   115–116, 247
Arroya del Rio, Carlos   109
Australia   3, 19, 49, 50, 89, 96–98
   forces in East Timor   181
Auer, Johannes   231
Austria   63, 146
Aviation, civil   12–13, 107–109, 180–181, 248–249
Azores   160–162

Backe, Herbert   69
Bananas   89
Barinen, Dudley   100
Bauxite   141
Beatty, A. Chester   41
Beef   90
Belgian Congo   98–101, 259
Belgium   83, 89
Ben Line   152–153
Biggs, Barton   18
Black markets   71–73, 77, 132–132, 190–191, 193
Bombay   4–5, 93
Bolivia   108, 110, 112, 117
Bondstedt, Eberhardt   106
Bonville, Louis   103
Brazil   106, 107, 110, 119
   foreign investment in   115
   rubber   56–57
   trade with Germany   114
Britain (*see Great Britain*)
Bristol Beaufort   97
British East Africa   239
British Empire   7–8, 36–37, 82, 83–84, 86–87, 265–266

British Guiana   86, 88, 108
British North Borneo Co   256–258
Broadcasting   2, 215–226, 269
Bulgaria   124
Burma Road   209–210
Busch, Germán   112

Caetani, Cora   126
Canada   5, 85, 104, 105, 95
Canned food   94, 171
Cape Verde   87
Cerjack-Boyna, Eric   112–113
Chad   91, 103
Changchun (*see Hsinking*)
Chiang Kai-shek   43, 199, 200
*Chicago Daily News*   268
Chile   52, 108, 119
China   20, 33, 34, 35, 76, 186,
    199–214
  Burma Road   209–210
  Chinese Maritime
    Customs   212–214
  Chungking, development
    of   204–206
  cotton mills   206–207
  Hainan   30, 177–178
  industrial capacity   34
  industrial relocation   204–209
  inflation   186, 199, 201–204
  Japanese occupied areas   169, 170,
    172, 173, 202–203
  Japanese surrender, post   243–244
  libraries, destruction of   211–212
  loans   25–26
  rice   200
  Shanghai, propaganda in   225–226
  tungsten   51–52
Chinese Maritime Customs   212–214
Chongqing (*see Chungking*)
Chrome   135–136
Chrysler   195
Chungking   174, 199, 201, 202,
    204–206, 213

Churchill, Winston   138, 160, 165
*Ciltvaira*   254–255
Coal   29–30, 93, 129–130, 141,
    196–197
Cocoa   88–89
Coffee   89
Colonial Office   7–8
Colonies   83–84, 89, 91
Co-Prosperity Sphere, Greater East
    Asian   168–186, 268–269
Corpening, Maxwell M   100
Copper   99
Copra   89
Croatia   123, 232
Cuba   19, 44
Currencies   182–184

de Gaulle, Charles   103, 217
de Bournat, Baron   105
Denmark   157, 259, 265
  food exports to Germany   64–69
Diamonds   54
Dupong, Pierre   230
Dutch East Indies (*see Netherlands
    Indies*)

East Timor   180–182
Ecuador   108, 109
Edgerton, David   viii
Eiichi, Baba   25
El Salvador   113–114, 116–117
Eritrea   120, 121, 123
Estigarribia, José Félix   111
Ethiopia (*see Abyssinia*)

Falk Corporation   34
Farben, I.G.   46
Faroe Islands   160, 267–268
Fisher, Charles A   171
Finland   155, 232, 251, 265
Fishing   68, 78–79, 176–177
Fleming, Alexander   252
Foianini, Dionisio   112

Food supplies   viii, 63–80, 94, 95, 96, 131–132, 165
Formosa   17, 30–31, 214
Forsyth, Bruce   3
France   71–73, 83, 101–105, 141, 150
  broadcasting   219
  postal restrictions   260
Franco, Francisco   90, 137, 138
French Cameroons   101
French Equatorial Africa   100, 101, 104
French Guiana   86, 88, 102
French India   101, 102–104
French Polynesia   102
Furhman, Arnulf   106

Gao, Bai   23–24
Ginsburgs, George   42
Germany   127, 128, 129, 130, 135–167
  aircraft production   140
  Argentina, investment in   247
  broadcasting   217, 220
  colonial legacy   81–82, 83–84
  Eastern Front   142
  financial controls   162–163
  food supply   63–71, 145
  and Greenland   157–159
  and Iceland   159–160
  industrial production   144–145
  influence in Latin America   106, 111–113
  investments in Argentina   115–116
  Labour situation   143–145, 148–149, 150
  'lebensraum'   39
  and Luxembourg   229–230
  oil production   45–48
  naval oil shortages   127
  and Norway   227–229
  railways   x, 164–165
  raw materials   39–48, 135–138
  shipping   151, 154

  supplies USSR   142
  and Sweden   154–156
  trade   114, 145–146
  whale oil   59–60
Ghana (see Gold Coast)
Gibraltar   138
Goebbels, Joseph   144, 226
Gold   2, 52–54
Gold Coast   44
Göring, Ernst   82
Göring, Hermann   46, 64, 142
Gould, Randall   208
Great Britain
  aircraft production   140–141
  and Azores   160–161
  defence spending   36
  Latin America, investment in   245–247
  motor industry   11–12
  oil   45–48
  pressure on Spain   136–138
  pressure on Turkey   135
  raw materials   xiii, 36–37, 38
  scientific achievements   37–38
  ship production   141
  smuggling from Switzerland   156–157
  supplying USSR   xiv
  wine imports   264–265
Greater Japan Women's Assn   187
Greece   231–232
Greenland   157–159
Guadeloupe   102

Hainan   30, 177–178
Haiti   55
Hankow Power & Light   30
Harris, Arthur   ix
Hassell, Ulrich von   67
Hastings, Max   xv, 190
Hay, John   58
Hayward, Joel   48
Henderson, Gavin   268

Henderson, Neville    1
Herman, Arthur    28
Hindustan Aircraft    94
Hitler, Bridget    221
Hitler, Adolf    1, 3, 43, 47, 84, 127,
    146, 221
Hong Kong    237–238
Hongkong & Shanghai Bank    243
Horses    ix, 75, 146
Hsinking    117, 168–169, 215
Huggins, Godfrey    90
Hull, Cordell    105
Hungary    232–234
Hunger Plan    69–71

Iceland    159–160, 232
India    19, 86, 103, 104, 92–95, 232
    war industries    92–93
Indian Army    82, 94
Indochina    78
Inflation    186, 199, 201–204,
    231–234
Iron ore    29–30, 49–50
Ishiguro, Tadaatsu    79
Italy    120–134, 232
    coal supplies    129–130
    colonial empire    120–123, 266
    lack of preparation    xi
    Latin American activities    106–
        107, 112, 113–114
    minerals    128–130
    oil    126–127
    railways    129
    rationing    130–132
    shortwave radio    216–217
    trade problems    123–124, 129

Jamaica    19, 89
Japan    14–35, 83, 108, 168–186, 187–
    198, 232
    activities in Latin
        America    116–118
    agriculture    75–76

black market    190–191, 193
budget problems    19
coal supplies    15, 32,    189,
    196–197
Co-Prosperity Sphere, Greater
    East Asian    168–186, 268–269
    currency control    182–184
    and East Timor    180–182
    electricity rationing    15
    fear of Stalin    11
    financial crisis 1927    23–24
    fishing industry    78–79, 176–177
    food supplies    viii, 79–80
    and Formosa    214
    military yen    182–184
    industrial organisation    18, 27–28,
        170–171, 173, 194, 241–242
    gold standard    24–25
    occupation of Manchuria    168–
        169, 174
    and Netherlands Indies    179–180
    oil    16–17, 48
    railway plan    196
    raw materials    28–32, 49–50
    relations with Germany    25,
        26–27
    rice supplies    192–193
    savings    189–192
    scientific dependence    18
    shipping    20–21, 31–32
    and Singapore    184–185
    and Southeast Asia    175–176
    surrender aftermath    240–244
    trade    19–20, 25
    whaling    59
    women in labour force    187–189
    zaibatsu    241–242
Joesten, Joachim    67
Joinovici, Joseph    72
Jones, Jesse H    57

Kailan Mines Administration    29–30,
    243

Kataoka, Naoharu   23
Knudsen, Helga   70
Knudsen, William S.   28
Krupp Armament Co   135
Kursk   143, 147

Labour control   90, 131, 143–144,
    187–189
Lanital   124–125
LATI (Linee Aeree Transcontinentali
    Italiane) 13, 106–107
Latin America   106–119
    advantages from war   2
    British investment in   245–247
    effects of war   118-119
    German airlines in   108–109
    German influence   110–113, 247
    Italian activities in   113–114
    Japanese activities in   116–118
    oil   48
    rubber   55, 56, 57
    U.S. blacklist   113
Latvia   253–255
League of Nations   x, xiv, 48, 60,
    74, 83, 96, 118, 143, 146, 149,
    250–252
Ley, Robert   xi-xiii
Liberia   55–56
Libya   120, 121, 122, 123, 266
Liberty ships   21–23
Lloyd, George (Lord)   89
Lockheed Aircraft Corp   ix
Luftwaffe   139, 140
Luxembourg   229–230

MacArthur, Douglas   241, 242
McSherry, Frank J   34
Macau   238
Madagascar   102
Maisky, Ivan   43
Malaya   77–78, 177, 236–237
Malcolm, Dougal   257
Manchukuo   116–117

Manganese   44
Martin, Norman   96
Martínez, Maximiliano
    Hernández   113, 117
Martinique   102
Meissner, Rudolf   106
Metlitzky, Bruno   111
Mexico   116, 119
Miklos, Bela   233
Mitsubishi   175
Møeller, John Christmas   66
Molotov, Vyacheslav   142
Monica, Sarah   vii
Montenegro   123
Moresco, Emanuel   83–84
Morínigo, Higinio   111
Morocco   87
Mountbatten, Louis   261
Mozambique   91, 92
Mozhaysk   147
Munch, Peter   67
Muselier, Émile   105
Mussolini, Benito   11, 121, 122, 124,
    126, 131, 133

Nakamoto, Hirako   vii, 188,
    197–198
Nathan, Edward   30
Nauru   98
Netherlands, the   45, 65, 83,
    149–150, 231
Netherlands Indies   83, 89,
    150–151, 175, 178–180,
    196–197,   234–235
New Caledonia   85, 86, 102
New Zealand   xv, 37, 98
Nickel   85
Nigeria   87
Norway   59–60, 65, 154, 227–229,
    265

Oil
    German needs   45 –48

Oil (*cont.*)
  Italian shortages   126–127
  Japanese shortages   15–16, 48
Olorunfemi, A.   xvi

Palm oil   88, 100
Pan American Airways   100, 161
Panama   112, 113
Panama Canal   108, 109, 112
Paraguay   111–112
Perón, Juan Domingo   247
Peru   6, 19, 108, 109, 117, 118
Petain, Phillippe   102, 258
Philippines   75–76, 235
*Picture Post*   4–5, 190, 256
Poland   65
Pontine Marshes   132, 134
Portugal   83, 92, 137, 146, 247, 259
  colonies   91
  East Timor   180–182
Postal   6, 12, 13, 104, 163–164,
    258–260
Propaganda   215–226
  German radio   217, 220
  Italy, shortwave   216–217
  Japanese   218–219
  Shanghai, propaganda
    centre   225–226
  U.S. broadcasting   220–224
Prora resort   xi–xiii, 263–264
Pu Yi, Henry   117, 168

Qantas   181

Raeder, Erich   127
Railways   x, 29, 81, 94, 101, 129,
    130, 144, 145, 154, 155, 164–165,
    168, 196, 197, 243
Reeves, John Pownall   238
Ribbentrop, Erich von   106
Ribbentrop, Joachim von   142
Rice   76, 77, 79, 192–193, 200, 243
Rice, Richard   27

Robilant, Edmondo di   107
Romania   44, 45, 47, 126
Roosevelt, Franklin D   17, 19, 28,
    39, 52, 58
Rubber   54–59, 100
Rügen   xi–xiii, 263–264
Russia (*see USSR*)
Ryckmans, Pierre   100

St Pierre et Miquelon   101, 104–105
Salazar, Antonio de   90, 161, 162
São Tomé & Príncipe   91, 92
Scindia Steam Navigation Co   94
Scott James, Anne   190–191
Shanghai   223–224
Shibusawa, Keizo   242
Shipping, merchant   94, 151–154,
    173, 234, 266–267
  Estonia   253
  Japanese shortages   20–21
  Latvia   252–255
  United States   21–23
Shipping, passenger   8–10
Silk   76
Singapore   77, 184–185, 235–236
Smith, Ben   264
Smith, N. Skene   27
Somaliland, British   87
Somaliland, French   102
Somaliland, Italian   120, 121, 123
Sommi, Luis V   116
South Africa   90, 98, 239
  gold   53
South-West Africa   81–82
Southern Rhodesia   90
Spain   146
  colonies   90
  tungsten exports   136–138
Speer, Albert   145, 147, 150
Starkenborgh Stachouwer, Jonkeer
    Tjarda van   178
Stalin, Joseph   43
Stalingrad   143, 144, 147

Stettinius, Edward, Jr   28
Stock exchanges   1, 162–163
Strength through Joy (KdF)   xi–xiii
Suez Canal   123
Sugar   95, 118
Svolos, Alexander   231
Swaziland   90–91
Sweden   74, 146, 154–156, 265
Switzerland   156–157
Syria   232

Taiwan (see Formosa)
Takahashi, Korekiyo   24–25
Tanganyika   87
Tata Steel   95
Textiles   19–20,   92–93, 94
Thailand   176
Tientsin   11
Tin   49, 58, 61, 85, 100
Topping, Helen   215
Trans-Siberian Railway   18, 43, 150, 151
Trepca Mines   41
Trinidad   88
Truman, Harry S.   22, 106, 107
Tungsten   51–52, 136–139
Turkey   135, 136, 146, 232

Ukraine   69–71
United Kingdom (see Great Britain)
United States   83
   blacklist Axis companies   113
   broadcasting   220–224
   China, forces in   243
   farmers   248–249
   industrial mobilisation   5–6, 28, 34, 262
   and Japanese in Latin America   117–118
   merchant shipping   21–23
   oil   48, 49

railways   197
rationing   75
raw materials   44, 50–52, 60–61
rubber   54–59
supplies USSR   165
tank production   195
tungsten   51–52
Uranium   50
Uruguay   110, 119, 232
U.S.S.R (Union of Soviet Socialist Republics also referred to as Soviet Union) 65, 164
   industrial resurgence   139–140
   supplies Germany   42–43, 142–143
   U.S. aid   165

Vargas, Getúlio   56
Vichy   73, 83, 91, 102, 104, 124, 140, 141, 163, 217, 219, 224, 258, 259
Vietnam (see Indochina)

Wagner Festival   3
Wallace, Henry A   222
Wang Ching-wei   172
War Manpower Commission   34
War materiel   95, 97, 139–140
Whaling   60–61
Whiteleather, Melvin K   73
Wilson, Edwin C   107
Winter, Hans von   113
Wolff, Emil   112
Women in industry   148–149, 187–189
World's Fair   10–11, 126, 251
Wu, Po-chau   205

Yasuba, Yasukichi   25
Yerrex, Lowell   109
Young, Hubert   236
Yugoslavia   41, 44, 264

# Other books by Robin Bromby

(Available on Amazon – in both e-book and paperback formats)

## German Raiders of the South Seas: The extraordinary true story of naval deception, daring and disguise 1914-1917

"The band had played *Watch on the Rhine*, the sailors left behind cheered, other men waved their hats, and women cried, as the *Emden* pulled out of Germany's China colony of Tsingtao for the last time" Thus begins a World War I epic that was, both literally and in spirit, a world away from the trenches and slaughter of France and Belgium. Were it not all real and true, it would make wonderful fiction: the *Emden* survivors from the battle with the Australian cruiser *Sydney* sailing a leaking copra schooner from the Cocos Islands to the East Indies, the *Wolf* sailing undetected in Allied waters for months, the captain of the *Seeadler*, von Luckner, sailing a small boat halfway across the Pacific to Fiji, and then later making a dramatic escape from a New Zealand prisoner of war camp.

## Australian Railways: Their Life & Times

Of all the torments suffered by the railway traveller in Australia none was so great as the break of gauge. At state borders, and within states, it was often impossible to complete a rail journey without changing trains when the track width changed. Western Australia and Queensland had gone for the narrow gauge, New South Wales opted for standard gauge, while Victoria went for the broad gauge. South Australia went one better: it opted for two gauges, narrow and broad. The nightmare of three different gauges, the daunting challenge of building railways across vast open spaces often with no water supplies, the follies of railway lines that were rarely used–all this is the saga of Australian railways.

## The Farming of Australia: A saga of backbreaking toil and tenacity

The story of farming in Australia has always been one of drama–two steps forward and then three steps back. From the days of the First Fleet through to modern farming of today, the industry has been beset by drought, floods, rabbit plagues, crop diseases, depressions, and disappearing markets.

www.ingramcontent.com/pod-product-compliance
Lightning Source LLC
Chambersburg PA
CBHW060542200326
41521CB00007B/450